Advanced Technologies and Societal Change

This series covers monographs, both authored and edited, conference proceedings and novel engineering literature related to technology enabled solutions in the area of Humanitarian and Philanthropic empowerment. The series includes sustainable humanitarian research outcomes, engineering innovations, material related to sustainable and lasting impact on health related challenges, technology enabled solutions to fight disasters, improve quality of life and underserved community solutions broadly. Impactful solutions fit to be scaled, research socially fit to be adopted and focused communities with rehabilitation related technological outcomes get a place in this series. The series also publishes proceedings from reputed engineering and technology conferences related to solar, water, electricity, green energy, social technological implications and agricultural solutions apart from humanitarian technology and human centric community based solutions.

Major areas of submission/contribution into this series include, but not limited to: Humanitarian solutions enabled by green technologies, medical technology, photonics technology, artificial intelligence and machine learning approaches, IOT based solutions, smart manufacturing solutions, smart industrial electronics, smart hospitals, robotics enabled engineering solutions, spectroscopy based solutions and sensor technology, smart villages, smart agriculture, any other technology fulfilling Humanitarian cause and low cost solutions to improve quality of life.

Vicente Garcia Diaz ·
Gloria Jeanette Rincón Aponte
Editors

Confidential Computing

Hardware Based Memory Protection

Springer

Editors
Vicente Garcia Diaz
University of Oviedo
Oviedo, Asturias, Spain

Gloria Jeanette Rincón Aponte
University Cooperativa de Colombia
Bogota, Colombia

ISSN 2191-6853 ISSN 2191-6861 (electronic)
Advanced Technologies and Societal Change
ISBN 978-981-19-3044-7 ISBN 978-981-19-3045-4 (eBook)
https://doi.org/10.1007/978-981-19-3045-4

This Springer imprint is published by the registered company Springer Nature Singapore Pte Ltd.
The registered company address is: 152 Beach Road, #21-01/04 Gateway East, Singapore 189721, Singapore

Preface

Confidential computing is a cloud computing technology that isolates sensitive data during processing in a protected CPU enclave. The contents of the enclave—the data being processed and the techniques used to process it—are visible and unknown to anyone or anything else, including the cloud provider.

As business leaders increasingly rely on public and hybrid cloud services, data privacy in the cloud is critical. The primary goal of confidential computing is to give leaders greater assurance that their data in the cloud is safe and secure and to encourage them to move more of their sensitive data and computing workloads to public cloud services.

This book highlights the three pillars of data security, viz. protecting data at rest, in transit, and in use. Protecting data at rest means using methods such as encryption or tokenization so that even if data is copied from a server or database an intruder cannot access the information. Protecting data in transit means making sure unauthorized parties cannot see information as it moves between servers and applications. There are well-established ways to provide both kinds of protection. Protecting data while in use, though, is especially tough because applications need to have data in the clear—not encrypted or otherwise protected—in order to compute. But that means malware can dump the contents of memory to steal information. It does not really matter if the data was encrypted on a server's hard drive if it is stolen while exposed in memory.

As computing moves to span multiple environments—from on-premises to public cloud to edge—organizations need protection controls that help safeguard sensitive IP and workload data wherever the data resides. Many organizations have declined to migrate some of their most sensitive applications to the cloud because of concerns about potential data exposure.

Confidential computing makes it possible for different organizations to combine data sets for analysis without accessing each other's data.

The book consists of 22 chapters, arranged on the basis of their approaches and contributions to the book and theme. The chapters of this textbook present key algorithms and theories that form the core of the technologies and applications concerned, consisting mainly of secure enclave technologies, adopting cloud computing, rise of confidential computing, confidentiality of data, components of a confidential computing architecture, confidential computing matters, IBM Cloud Data Shield vs. Microsoft Azure Confidential Computing, isolating the software and data from the underlying infrastructure, hardware-level encryption, public clouds, secure and private analytics, blockchain, paradigm shift for data security in the cloud, benefits of confidential computing.

Oviedo, Spain — Vicente Garcia Diaz
Bogota, Colombia — Gloria Jeanette Rincón Aponte

Contents

Chapter 1
Design and Implementation of Mobile Jammer for High Security System

M. Nagaraju Naik, S. Nehasree, N. Sai Pramodha, and G. Mounika

Introduction

Jamming devices work by transmitting a signal on the same frequency as the mobile phone and at a high enough power to cause the two signals to overlap and cancel each other out. Since cell phones are programmed to increase power when they encounter low-level interference, the jammer must be able to detect and balance the phone's increased power [1, 2]. Cell phones are full-duplex devices, which means they use two frequencies at the same time: one for communicating and one for listening. Some jammers only block one of the mobile phone frequencies, but this effectively blocks both. Since it can only receive one of the frequencies, the phone is fooled into believing there is no coverage [3]. Less sophisticated devices block only one group of frequencies, while more sophisticated jammers will block several types of networks at once, preventing dual-mode or tri-mode phones from switching between various network types to reach an accessible signal [4].

Literature Review

It used to be difficult to catch copying students in the examination hall because teachers had issues with it, and it was a more difficult task. We are now using CCTVs, etc.

M. Nagaraju Naik (✉) · S. Nehasree · N. Sai Pramodha · G. Mounika
Department of Electronics and Communication Engineering, CMR College of Engineering & Technology (Autonomous), Kandlakoya, Hyderabad, India
e-mail: nagarajunaik1976@cmrcet.org

V. Garcia Diaz and G. J. Rincón Aponte (eds.), *Confidential Computing*,
Advanced Technologies and Societal Change,
https://doi.org/10.1007/978-981-19-3045-4_1

1. CCTV

   Closed circuit television (CCTV): In examination halls, cameras are used to detect cheaters and discourage unfounded lawsuits against teachers or invigilators. And these are placed in such a way that only one CCTV per room and one person manually watching is necessary to prevent cheating in the examination hall; however, manual work has the option of recording, which the squad will review later [5, 6]. The following are some of the drawbacks of this CCTV system:

   (i) If there is no backup power (power outage)
   (ii) (Distraction occurs) when we attach a magnet to it
   (iii) It is unable to identify anyone who is using a cell phone.

2. Manual Surveillance

   One individual can manually search the students for chits in manual surveillance, but this is a more time-consuming and exhausting operation. This procedure [2] is ineffective as compared to CCTV because it has a high risk of escaping because verifying everybody is a difficult job, and most people are checked for the sake of checking [1, 7].

   The following are some of the drawbacks of this CCTV system:

   i. Inadequate checking (like palm, etc.)
   ii. Tough process
   iii. There is a good chance you will be able to get away.

3. Spoofing

   This form of jamming causes the phone to shut down on its own. Since the spoofing mechanism first detects every cell phone in a specific location, then the system senses the signal to disable the mobile phone, and this type of jamming is extremely difficult to implement. Certain techniques can detect the presence of a nearby mobile phone and send a message instructing the user or individual to enable silent mode or turn off mode [8, 9].

4. SMS Spoofing

   This method is used because it is compatible with the majority of cell phones. This has both legitimate and illegitimate uses (setting the company name from which messages are sent, setting your own mobile number such as impersonating another person, company) [10, 11].

Existing System

TV remote controls, air conditioner remote controls, toy remote controls, car remote controls, and so on are all commonplace in our everyday lives. Not all remote controls use the same frequency bands, which include 315, 330, 390, 433, 868 MHz, and a variety of others. And there's no question that the remote control jammer's design is to disable the remote control's signals. On the one side, remote controls such as the TV remote control and the air conditioner remote control can provide us with a great deal of convenience; however, the remote control can also control a variety of other devices [12].

Adjustable mobile jammer is a full jamming solution that can provide mobile jamming 24 h a day, seven days a week. It is a simple system that can be turned on to block cell phone signals in conference rooms, hospitals, libraries, recording studios, and other places to prevent disruption and give you a break from phone calls and texts, allowing you to focus on your job. It is an effective jammer that helps to ensure complete silence [13]. Adjustable mobile phone jammers are the most advanced form of jammer currently available on the market. This is due to the fact that they are fixed devices with customizable settings, making it easy to keep cell phone chatter and distractions out of vast areas designated as mobile free zones. These mobile jammers are also very useful in schools and tech companies because they allow us to change locations and jam where we expect silence while excluding places where we do not need to jam signals [14, 15].

A portable mobile jammer is small enough to fit in your pocket. To build a quiet zone around you or prevent information leaking in sensitive areas, choose from cell phone only or combination versions that include GPS, or even our most common model, the cell phone blocker mini. The portable cell phone signal blocker is small in size but strong in function [16, 17]. A compact one's blocking radius is no different from that of larger ones. Its working range is approximately 10–100 m. People are increasingly embracing portable jammers due to their comfort and benefits. For example, many students in senior and high school already own cell phones, and many choose to play with them instead of paying attention in class [18, 19]. To increase class performance, the teacher should use a portable cell phone jammer to block cell phone signals. Some of the disadvantages are not in easy-to-use, more expensive, and complex design.

Implementation

Requirement Analysis

Mobile jammer mainly consists of.

(i) Hex 74LS04-Inverter
(ii) 9v battery with snap
(iii) Tuning capacitor
(iv) Antenna
(v) PCB layout
(vi) Connecting wires

Hardware Specifications

Hex 74LS04-Inverter

Hex 74LS04-Inverter: This is a standard inverter that takes a signal as input and outputs a blocked signal. It is similar to a hex inverter chip in that it has six separate inverter gates. Each gate has a single input and output. The output becomes high when the applied input is low and vice versa. It is made up of 14 pins:

Pin 1, 1A Input Gate 1

Pin 2,1Y Output Gate 1

Pin 3,2A Input Gate 2

Pin 4,2Y Output Gate 2

Pin 5,3A Input Gate 3

Pin 6,3Y Output Gate 3

Pin 7, Ground

Pin 8,4Y Output Gate 4

Pin 9, 4A Input Gate 4

Pin 10,5Y Output Gate 5

Pin 11,5A Input Gate 5

Pin 12,6Y Output Gate 6

Pin 13,6A Input Gate 6

Pin 14, Supply(vcc)

9v Battery with Snap

This is used to supply power to the entire system, and the snap connects the entire set-up to the battery. We attached the battery in this project so that the cathode is connected to the 14th pin, which is the voltage pin on the inverter, and the anode is connected to the 7th pin, which is the ground pin on the inverter.

Tuning Capacitor

There are two kinds of variable capacitors. One of the forms of basic components is the tuning capacitor (variable capacitor). Tuning condensers are another name for tuning capacitors. It has three terminals. The variable capacitor is a capacitor whose capacitance can be adjusted mechanically or electronically on a regular basis. Blocks are used in the construction of these. This capacitor's capacitance can be adjusted to a specific range depending on the use. In L/C circuits, variable capacitors are often used to set the resonance frequency. In our project, one terminal is attached to the inverter, the other to the antenna, and the third is left unconnected.

Antenna

The interface between signals propagating or travelling through space and electric currents moving through metal conductors is the antenna, which is used in conjunction with a transmitter or receiver. It is a specialized transducer that transforms radio frequency (RF) fields into alternating current (AC) or the other way around. The receiving antenna and the transmitting antenna are the two forms. These are mostly used in computer and Internet wireless applications; the most popular type of antenna is a dish antenna, which is used for satellite communications. In our project, we wired the antenna so that the end point is connected to the tuning capacitor's negative terminal.

PCB Layout

A printed circuit board, also known as a printed wiring board, is a form of circuit board. It is primarily used in the manufacture of electronic devices. This PCB layout serves two key purposes: (1) providing a location to place the components and (2) providing a means of connecting the components electrically.

Connecting Wires

This allows current to flow from one point on a circuit to another. It has the ability to draw power and completes its mission.

Methodology

The Hex 74LS04-Inverter is a 14 pin IC. Internally wired three input pins and three output pins are individually connected to the antenna via an inverter, and tuning capacitor is connected in series with input, output pins, and antenna. A 9v battery is connected across the VCC supply pin and the ground pin to change the frequency from 0 to 108 MHz. Although it is an inverter, it converts 0 to 1 and 1 to 0, so when power is applied, the signal is reversed, resulting in signal jamming.

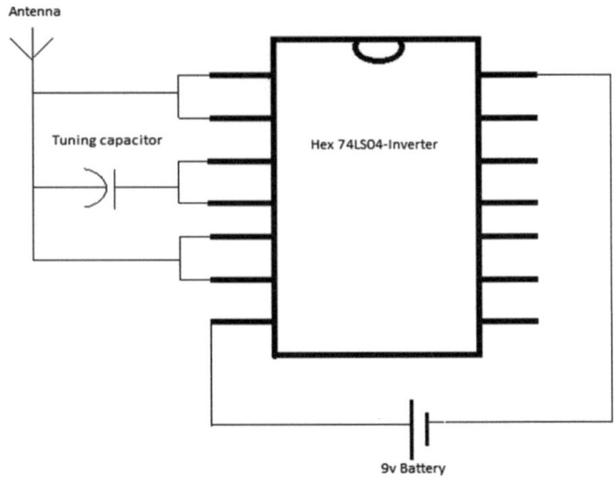

Fig. 1.1 Pin diagram of Hex 74LS04-Inverter circuit diagram

Circuit Diagram and Description

Figure 1.1 is a circuit diagram of mobile jammer. The inverter is held on the PCB configuration, and the 7th and 14th pins are used for battery connections, with the remaining pins connected internally. The second input and output terminals are attached to a tuning capacitor. The antenna's first, second, and third input–output terminals are all attached to the antenna's end. When the supply is given, the signal is blocked by this jammer.

System Design

See Fig. 1.2

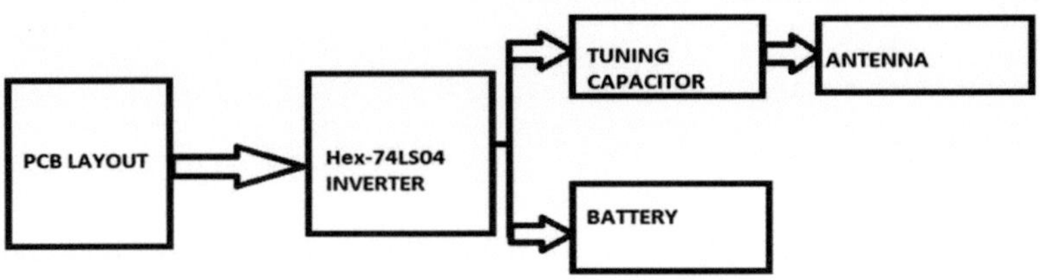

Fig. 1.2 Implementation model system design

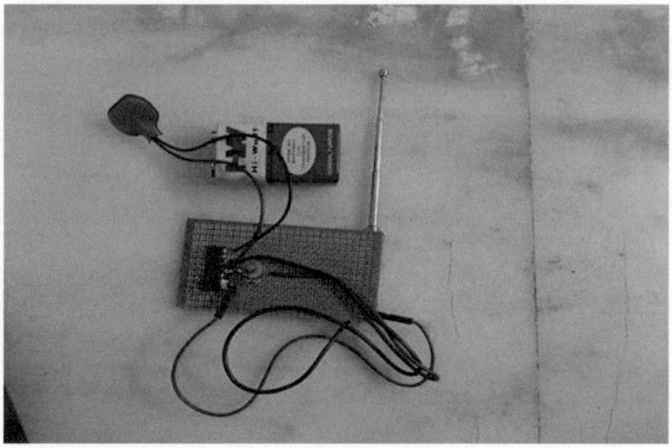

Fig. 1.3 Prototype

Results

We created a sophisticated prototype mobile jammer. This prototype is low cost and efficient, and it is designed primarily for use in schools, colleges, offices, and other private workplaces. Prototype signals developed can block up to 40 m in conference rooms and are tuned at a frequency range of 1800–1885 MHz. We experimented with several frequencies, and a 108 MHz variable capacitor was utilized to set the frequency for 40 m. To further expand the signal blocking area, frequency tuning with appropriate capacitors was necessary (Fig. 1.3).

Conclusion

The objective of a mobile jammer is to control a certain frequency for signal blockage. Conferences and other private work locations required more concentrate on their work, but mobile phones were making more noise during important meetings, thus jammers may play an important part in creating a cool environment. As a result, we created a low-cost and efficient prototype. Different frequencies were tested, and the frequency was set to 1800–1885 MHz. The mobile jammer is an excellent choice for a secure system.

References

1. Wankhede, H.K., Wankhede, K., Jethwa, M.A., et al.: Blocking the mobile phones signals with the help of jammer. Int. J. Mod. Trends Eng. Res. **2**(7), 110–114 (2015)
2. Naresh, P., Babu, P.R., Satyaswathi, K.: Mobile phone signal jammer for GSM, CDMA with pre-scheduled time duration using ARM7. Int. J. Sci. Eng. Technol. Res. **2**(9), 1781–1784 (2013)

3. Abdul-Rahman, A.S.H., Mohammad, A.N.R.: Dual band mobile jammer for GSM 900 & GSM 1800. Technical Report, access on Maret: Jordan University of Science and Technology (2013)
4. Singh, S., Kaur, R.: Blocking the phone signals with the help of mobile jammer. Int. J. Innov. Res. Comput. Commun. Eng. **4**(3), 4225–4231 (2016)
5. Australian Communications Authority: ACA Report, Mobile phone jammers (2003)
6. Mishra, N.K.: Development of GSM-900 mobile jammer: an approach to overcome existing limitation of jammer. In: IEEE Fifth Conference on Wireless Communication and Sensor Networks (WCSN), pp. 1–4 (2009)
7. Punal, O., Aguiar, A., Gross, J.: In VANETs we trust?: characterizing RF jamming in vehicular networks. In: Proceedings of the Ninth ACM International Workshop on Vehicular Inter-Networking, Systems, and Applications, pp. 83–92. ACM (2012)
8. Kanojiya, V.U., Yadav, J.B.: Implementing mobile jammer in automobiles. Int. J. Adv. Res. Comput. Sci. Manage. Stud. **3**(5), pp. 508–512 (2015). ISSN:2321-7782 (Online)
9. Olumide, O.O., Jimoh, O.O., Olaide, L.A.: Design and development of mobile phone jammer. Am. J. Eng. Res. **5**(2), 71–76. e-ISSN: 2320-0847, p-ISSN: 2320-0936
10. Patel, I., Shigli, A., Sripathi Raja, V., Kulkarni, R.: Intelligent mobile signal jammer. Asian J. Comput. Sci. Eng. **2**(5), 01–06 (2017)
11. Zhang, L., Wang, H., Li, T.: Anti jamming message-driven frequency hopping-part-1: System design. IEEE Trans Wirel Commun **12**(1), 70–79 (2012)
12. Razazadesh, N., Shafai, L.: A compact antenna for GPS anti-jamming in airborne applications. IEEE Access **7**, 154253–154259 (2019)
13. Wang, B., Wu, Y., Liu, K.R., Clancy, T.C.: An anti-jamming stochastic game for cognitive radio networks. IEEE J Selected Areas Commun 29(4), 877–889 (2011)
14. Bhavani, M., Narayana, V.A., Sreevani, G.: A novel approach for detecting near-duplicate web documents by considering images, text, size of the document and domain. In: Lecture Notes in Electrical Engineering, vol. 398, pp. 1355–1366 (2021)
15. Premalatha, B., Srikanth, G., Abhilash, G.: Design and analysis of multi band notched MIMO antenna for portable UWB applications. Wirel. Personal Commun. **118**(2), 1697–1708 (2021)
16. Rani, P., Mishra, A.R., Ansari, M.D., Ali, J.: Assessment of performance of telecom service providers using intuitionistic fuzzy grey relational analysis framework (IF-GRA). Soft Comput. **25**(3), 1983–1993 (2021)
17. Rashid, E., Prakash, M., Ansari, M.D., Gunjan, V.K.: Formalizing open source software quality assurance model by identifying common features from open source software projects. In: Lecture Notes in Electrical Engineering, vol. 698, pp. 1375–1384 (2021)
18. Debnath, S., Talukdar, F.A., Islam, M.: Combination of contrast enhanced fuzzy c-means (CEFCM) clustering and pixel based voxel mapping technique (PBVMT) for three dimensional brain tumour detection. J Ambient Intell. Human. Comput. **12**(2), 2421–2433 (2021)
19. Merugu, S., Reddy, M.C.S., Goyal, E., Piplani, L.: Text message classification using supervised machine learning algorithms. In: Lecture Notes in Electrical Engineering, vol. 500, pp. 141–150 (2019)

Chapter 2
Dual Security Based Attendence System by Using Face Recognition and RFID with GSM

V. Hemalatha, B. Premalatha, and K. Kiran Kumar

Introduction

Face based totally recognition of people can be very beneficial to investigate their identity. Many papers have actually been endorsed associated with RFID in addition to additionally finger print based totally presence device. We are incorporating the face on-line recognition techniques in addition to subsequently suggesting a model in an effort to no longer most effective be sensible for participation recording and additionally tracking but likewise it is going to enhance protection and protection and safety. We are providing a maker wherein the Student statistics is produced inside the challenge commercial enterprise business enterprise or some distinct institutional database. The photo may be launched right into the truths supply as nicely. Making use of appropriate face advice solution, the face matching is executed with the Pupil on the identical time as he's taking component inside the university that is stuck the usage of the electronic cameras. This motion acts as very first layer of protection for the touchy places. If healthy of the face does no longer appear after that alarm obtains punctual & safety guards will do some thing favorable concerning it [1, 2]. If the face of the person getting into college suits with the without a doubt available database. Every unmarried time he asks for to swipe his ID card each for acquiring gets right of entry to proper into the college in addition to additionally as leave. With this form of gadget we are able to genuinely limit the unapproved utilization ID betting playing

V. Hemalatha (✉) · B. Premalatha
Department of ECE, CMR College of Engineering & Technology, Hyderabad, Telangana, India

B. Premalatha
e-mail: bpremalatha@cmrcet.org

K. Kiran Kumar
Department of Computer Science and Engineering, Koneru Lakshmaiah Education Foundation, Guntur, Andhra Pradesh, India
e-mail: kiran5434@kluniversity.in

V. Garcia Diaz and G. J. Rincón Aponte (eds.), *Confidential Computing*,
Advanced Technologies and Societal Change,
https://doi.org/10.1007/978-981-19-3045-4_2

cards as we're using face acknowledgment present day technology further within the advised gizmo. Dual or multi people the front with singular gets admission to card also can be stop [3, 4].

Face credibility advancement is a incredible deal higher than various specific biometric broadly speaking primarily based acknowledgment techniques like fingerprint, palm-print, iris due to its non-contact fashion. Recommendation methods the use of face on-line recognition can moreover grow to be knowledgeable approximately a personality from a variety, without a hook up with or interaction with personality. The face acknowledgment strategies are presently used in social networks internet web site like on the flight terminals, educate stations. Then, at crook pastime checks [5, 6]. Face attraction approach can moreover be used in scoundrel enjoyment hobby reviews; the captured photo is probably saved in a information source, even more to might be applied to set up a personality. Social area makes use of the face acknowledgment method for automating the device of identifying people. For face reputation we require huge dataset and also hard capabilities to understand a character in all situations like adjustment of illumination, age, pose, and so forth. Current seems into software program there might be betterment in facial acknowledgment frameworks. In the very last 10 years there is big rise in acknowledgment techniques [7, 8].

But underneath we are the usage of dual safety innovation for operation and additionally create the protection; the usage of RFID & Face popularity-primarily based technique with IOT platform (Fig. 2.1).

Main Objective:

Whenever the Student goes into the college, he's required to swipe his/her launched Student ID card from the university. The video clip virtual needs to be put close to the doorway way. When the Pupil swipes ID card variety and ID quantity does currently not match with the truths supply rather exams out the stats. When RFID site traffic

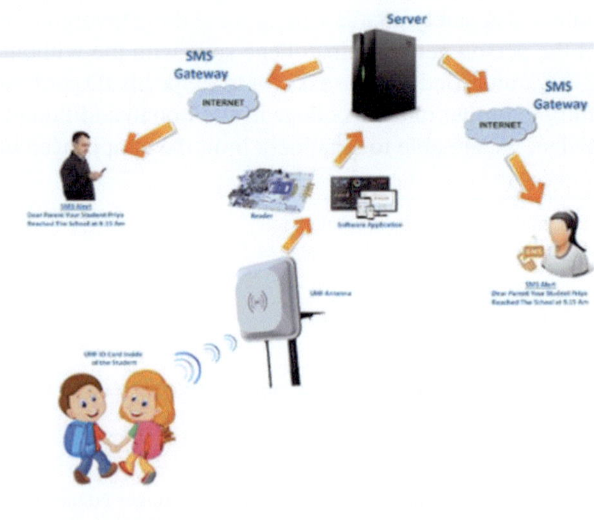

Fig. 2.1 Proposed operation model

gather files from card, and then maximum in all likelihood to 2nd protection as well as security, i.e., Face on-line reputation cutting-edge era with the useful supply of making use of USB video digital.

Related Study

Many obligations were carried out the use of Arduino for reinforcing revolutionary participation equipment. Considered that scholars from numerous the world over regions, typically, from setting up mixed with underdeveloped are short growing [3–5], the decision for in a comparable way to enchantment of automatic tool based high-quality beauty paintings are setting up. For example, IR module-based totally system for present day have a have a take a look at area has in reality evidently been completed in countless areas [9]. Besides, Bluetooth factor involvement device thru technique of which encompass the cellular smart tool devices has virtually clearly been taken advantage of via data colleges [10]. One more interest that we located extra concerning during our test is GSM based totally truly automated university vicinity system [11]. Formerly, RFID based totally absolutely look at area device has in reality basically been supported the use of raspberry pi in addition to ATmega32 increase package deal [12]. Nonetheless, raspberry pi or ATmega32 upward thrust bundle good deal does no extra make our challenge fee effective [13]. We are misting probable to make this venture for the academic institutes; as a given up result aside from the safety coverage of precision and credibility, charge common modern-day everyday overall performance is moreover a huge problem to us. In this paper, we subsidized a presence machine in particular primarily based upon Face Recommendation on the facet of Verified the stats with manner of RFID moreover to in the end defend info via identifying face, figuring out identity range, the front further to leave time by using Real time Clock (RTC) detail [14–16]. This records is probably logged the usage of the use of SD card or the usage of sending it to the net thru the use of an Ethernet guard, based absolutely actually honestly upon clients' requirement [17]. The following via elements of the paper are placed in an order because of this. Area 2 makes a specialized of the layout of our device that we've got was provided currently influenced. Area three represents the circuit format with right situations in addition to moreover without a doubt simply how we had honestly completed our gadget of this assignment. Last but no longer least, Location four is the honestly last precept in which we moreover went over discussing the future amount.

Existing System

The contemporary attendance device calls for trainees to manually signal the sheet on every occasion they participate in a class. This includes the even extra time eaten by using the trainees to discover their call on sheet, a few university college students would likely mistakenly authorize an extra pupil's name as well as the regularly sheet might likely achieve misplaced. For warding off the sheet problem, we made use of RFID innovation, RFID card document the presence with the aid of way of flashing their card and moreover hold all information but from time to time RFID card may additionally obtained misplaced [18]. Managing attendance is probably a vital record-keeping venture in any business enterprise. There are various strategies of keeping the participation from manual device in which participation is mentioned in sheets to computerized participation on the likes of biometrics. The blend of frequency Identification (RFID) with fingerprint biometric modern generation changed into to enhance the safety and protection degree and balance of the files. The made device no longer simply makes the system fashion less complicated however likewise complements the productiveness of the agency every with reference to man electricity as well as time. The machine does now not simply simplify the approach of taking attendance but decreases mistake and lets in faster verification of student presence, all with minimum human communication. This approach will assist the government manipulate the participation device in an additional prepared, dependable and moreover time preserving manner.

Proposed System

The aim of this system is to broaden a cordless device to find out and additionally hold the presence of a student further to situate a pupil. In this venture we taken student presence with the aid of using two level protection and protection, one is face acknowledgment and an additional one is RFID card acknowledgment (ID playing cards), each time the scholar no longer comprise college/college after that at once the facts is send to parents and in addition to database(server). In this method we're supply 2 stage safety via manner of this protection the student maximum definitely need to carry identity playing cards, the attendance is stored through server as well as message ship out to precise authority individual, based upon the situation they can do something effective approximately it rapid.

As talked about over only for valid state of affairs, Presence is mentioned for the Pupil with the get proper of access to time & day details & front gateway is opened for the Student get right of entry to. If the circumstance is for a valid identity card yet the face isn't matching with that stated of nowadays in the database for that certain man or woman. Safety is straight away knowledgeable with this & because of this trespasser will genuinely be captured with the resource of the safety humans. Thus, to videotape the participation Student desires to revel in three stages of affirmation which incorporates Legitimate Student ID card matched after that determined matched. When these parameters are surpassed after that truly the Existence will really be tape-recorded with the front records.

Materials and Methods

Currently days we find RFID based attendance gadget. In the proposed device we've protected the blessings of RFID with the face acknowledgment strategies, for this reason supplying a two-layer complete proof protection machine for any sort of sensitive establishments (Fig. 2.2).

RFID reader:

RFID is a word for Super high frequency Recognition. RFID (radio frequency recognition) is a contemporary-day era that includes the usage of electromagnetic or electrostatic coupling in the super high frequency (RF) segment of the electromagnetic spectrum to distinctively discover an object, animal, or individual. RFID is moving into elevating usage in industry as a choice to bench code. The benefit of RFID is that it does not require direct touch or line-of-sight scanning. An RFID gadget includes 3 components: an antenna and transceiver (typically combined into one visitor) and also a transponder (the tag). The antenna makes use of super high frequency waves to ship a signal that activates the transponder. When activated, the tag transfers data returned to the antenna. The statistics is applied to alert a programmable reasoning controller that and movement must stand up. The interest can be as essential as elevating an accessibility the front or as complex as interfacing with a statistics supply to perform a monetary buy.

GSM Module:

A GSM modem or GSM difficulty is a hardware device that makes use of GSM cellular Smartphone technology to provide a facts hyperlink to a much-flung community. From the view of the cell Smartphone community, they'll be essentially much like a common mobile Smartphone, along facet the selection for a SIM to discover them to the network. GSM modems usually offer TTL-diploma serial interfaces to their host. They are commonly made use of as part of an embedded system.

USB Camera:

An electronic cam is a video digital camera that captures pix in virtual memory. The majority of cameras generated nowadays are digital, considerably converting folks that capture images on photo film. While there are though dedicated virtual

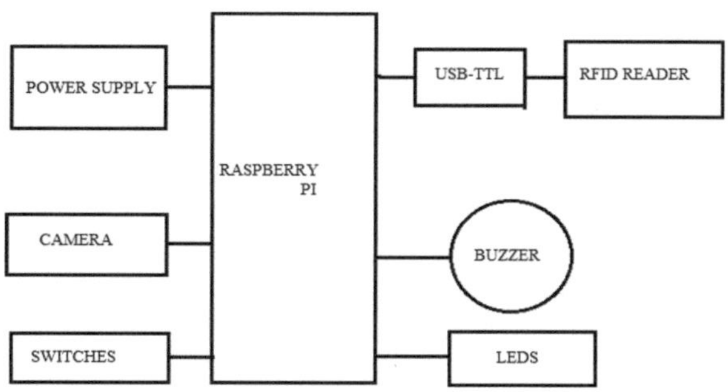

Fig. 2.2 Proposed block diagram

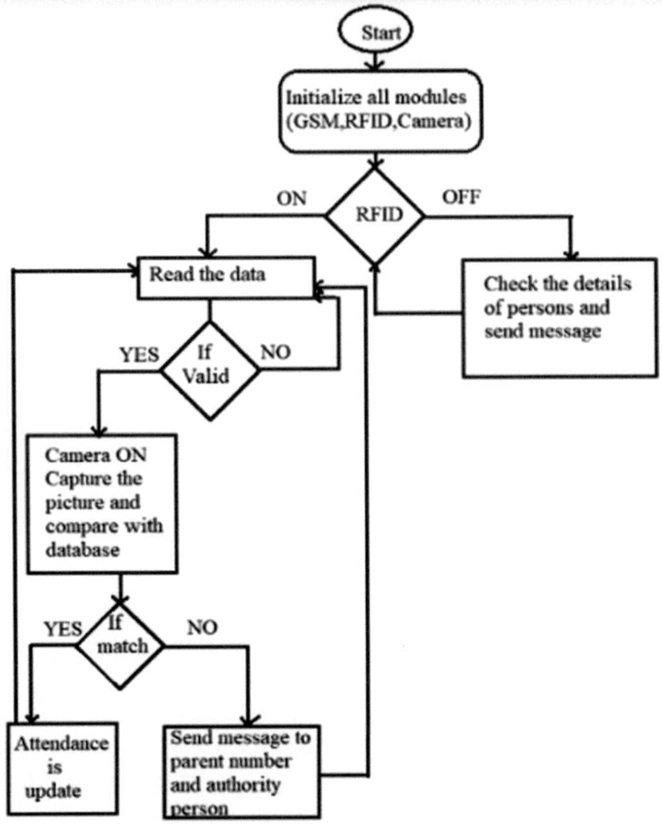

Fig. 2.3 Flow chart of the working

cams, many greater cameras are presently included right into cellular phones like clever telephones, that could, amongst several various special purposes, use their video cameras to provoke actual-time video-telephony and moreover immediately edit and additionally post photos to others. High-surrender, excessive-def specialized video cameras are however commonly used by experts and those that choice to take higher-excellent images (Fig. 2.3).

Working Methodology with Results Explanation

In the complying with place, we're misting probable to observe about exactly how we carried out the system of our counseled challenge Attendance System based totally on Face Acknowledgment and Confirmation via RFID. From Number provide a reason behind the hardware packages (Fig. 2.4).

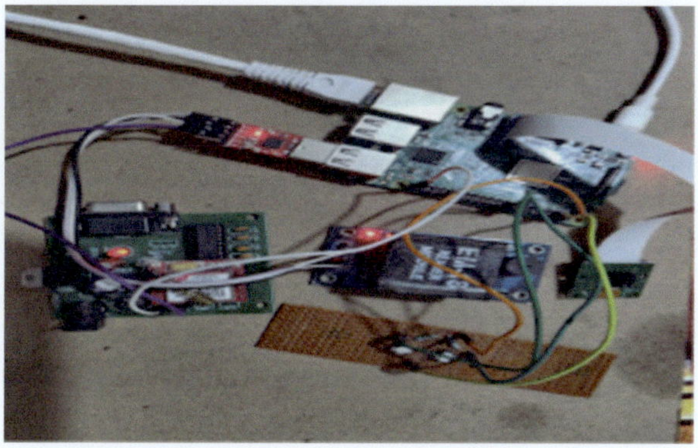

Fig. 2.4 Hardware kit with RFID reader

Figure is revealing the smooth circuit format of our project. Listed beneath we will see that the LCD is installed to the bread board. Likewise, a LED, a RTC are associated. Every considered one of them blended related to Raspberry Pi. Along with the Raspberry Pi is connected to the computer from wherein it is a good deal going to get electric electricity at the side of run the whole records. The LCD in addition to moreover LED is of entirety outcome proper right here. There is a pot additionally which suits because the evaluation controller of the LCD show. We can set up the agreement via using it each time we want to try this (Fig. 2.5).

Number five is revealing us that exactly simply exactly how we're misting probably to update our laptop system pc laptop pc registry or display. Usually, at the identical time as us going the deliver the center to the realities useful resource the tool will in reality request the selection of the student and furthermore in addition the ID. We want to go into those stats effectively.

Figure 2.6 is displaying us that what we're successful of really do after the listing beneath motion. After get in the name similarly to ID this device will basically request for the photo of the scholar. He/she wants to stand in the front of the cam collectively with the internet cam will in truth take 20 snaps concurrently in addition to hold it in

Fig. 2.5 Camera connected for face recognition

Fig. 2.6 GSM module interfacing with Raspberry PI

its records supply. Something we preference to remember that for accuracy we want to transfer our face particularly factor of view and there need to suffice modest.

We can see that the tool has effectively diagnosed each trainees whose info and moreover photo became participated within the database. Whenever a pupil is misting most in all likelihood to stand in the front of the video clip digital internet cam if he's certified it's going to truly show his ID right into the rectangular block and moreover his lifestyles will certainly be long past to the top database robotically. Afterwards he can without issue take part inside the majesty location through making use of passing the IR elements. The above numbers lay out every segment specifically precisely how the device attributes. Face is identified after being matched with saved stats. As rapid because of the truth the face is unique, it widely known the face with 172 identity complete variety. Complying with phase is to show ID card to the traveler, LCD show spherical all once more suggests his data. After that as he goes across the pinnacle IR deciding on up machine to go into the sophistication, issue variety will in reality increase. This counted range will actually exist on LCD. As the students go away the route vicinity going across some other IR sensor; preserve in thoughts declines as well as is finished on LCD. We can perform each device, address acknowledgment and additionally Recognition via the use of RFID to merge in a single machine similarly to furthermore encompass a storage device for our venture. Due to absence of time, we can't include the attribute but (Figs. 2.7 and 2.8).

Conclusion

The fashion and also execution of the Participation System primarily based entirely mostly on Face Acknowledgment similarly to Verification via RFID that have grow to be our purpose in addition to additionally objective of the paper at the start finishes with an accomplishment as each element works as favored. There it does without any type of kind of mentioning that our endorsed model has the capability to put off the manual presence device as it's dependable in addition to hassle-free. Our version is a

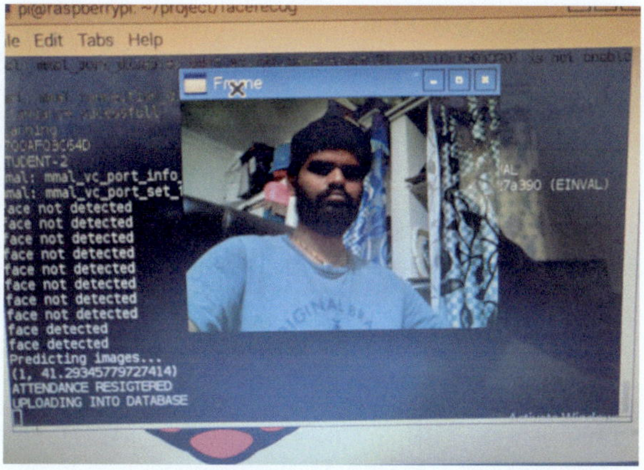

Fig. 2.7 Camera OUTPUT in PC

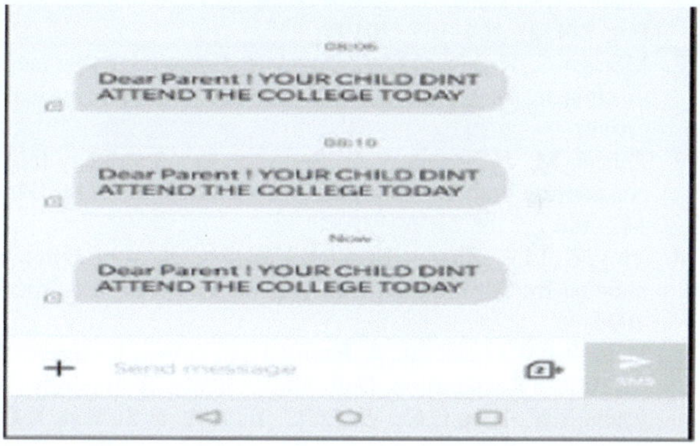

Fig. 2.8 OUTPUT results from GSM to parent mobile

horrible lot higher person terrific and also moreover it offers the most actual further too prepared facts. And with surely more than one adjustment we're capable of use our device in any shape of blanketed facilities.

References

1. Amiyana, N., Alias, M.: Attendance and access control system using RFID system. Undergraduate Thesis, Faculty of Electronic and Computer Engineering, Universiti Teknikal Malaysia Melaka, pp.1–24 (2011)
2. Hapani, S., et al.: Automated attendance system using image processing. In: 2018 Fourth International Conference on Computing Communication Control and Automation (ICCUBEA). IEEE, 2018
3. Akbar, Md.S., et al.: Face recognition and RFID verified attendance system. In: 2018 International Conference on Computing, Electronics & Communications Engineering (iCCECE). IEEE, 2018

4. Okokpujie, K.O., et al.: Design and implementation of a student attendance system using iris biometric recognition. In: 2017 International Conference on Computational Science and Computational Intelligence (CSCI). IEEE, 2017
5. Rathod, H., et al.: Automated attendance system using machine learning approach. In: 2017 International Conference on Nascent Technologies in Engineering (ICNTE). IEEE, 2017
6. Siswanto, A.R.S., Nugroho, A.S., Galinium, M.:. Implementation of face recognition algorithm for biometrics-based time attendance system. In: 2014 International Conference on ICT For Smart Society (ICISS). IEEE, 2014
7. Lukas, S., et al.: Student attendance system in classroom using face recognition technique. In: 2016 International Conference on Information and Communication Technology Convergence (ICTC). IEEE, 2016
8. Zhang, Z.N., John, Y.W.: Inducement analysis in facial expression recognition. In: 8th International Conference on Signal Processing, pp. 1654–1657. IEEE Press, 2006
9. Nagraju Naik, M., Venkata Subba Reddy, K.: Spatial correlation based contrast enhancement for retinal images. Int. J. Innov. Technol. Explor. Eng. **8**(452), 130–134 (2019)
10. Papageorgiou, C.P., Oren, M.: A general framework for object detection. Comput. Vision, IEEE Int. Conf. **0**, 555–562 (1998)
11. Shaik, A.S., Usha, S.: Sensor based garbage disposal system. Int. J. Innov. Technol. Explor. Eng. **8**(452), 164–167 (2019)
12. Merugu, S., Reddy, M.C.S., Goyal, E., Piplani, L.: Text message classification using supervised machine learning algorithms. In: Lecture Notes in Electrical Engineering, vol. 500, pp. 141–150 (2019)
13. Zaheer, M., Narayana, V.A., Sreevani, G.: A strategy for near-deduplication web documents considering both domain & size of the document. Int. J. Innov. Technol. Explor. Eng. **8**(452), 141–146
14. Ying, Z., Li, J., Zhang., Samuel, Y.W.: Facial expression recognition based on classifier combinations. In: 8th International Conference on Signal Processing, pp.1628–1632. IEEE Press, 2006
15. Duin, R.P.W., Juszczak, P., Paclik, P., Pekalska, E., Ridder, D., Tax, D.M.J.: A Matlab Toolbox for Pattern Recognition. Delft University of Technology (2004)
16. Chen, Y.N., Han, C.C., Wang, C. T., Jeng, B.S., Fan, K.C.: A CNNbased face detector with a simple feature map and a coarse-to-fine classifier—Withdrawn. IEEE Trans. Pattern Anal. Mach. Intell. **99**(l–l)
17. Swapna Rani, T., Bhardwaj, K.K.: Comparative analysis of 8-Bit ALU in 90 and 45 nm technologies using GDI technique. In: Lecture Notes in Networks and Systems, vol. 65, pp. 403–408 (2019)
18. Torralba, A., Murphy, K.P., Freeman, W.T.: Sharing visual features for multiclass and multiview object detection. IEEE Trans. Pattern Anal. Mach. Intell. **29**(5), 854–869 (2007)

Chapter 3
Disaster Analysis on Government Data

T. Nirmala, Shiva Akshith Kumar, P. Rithvik Rao, P. Raviteja Reddy, and T. Poojitha

Introduction

Disasters are caused due to natural and man-made calamities. There are a lot of cases in the world currently the order of immediate situation that existed in our regular basis in USA dataset taken for disaster management. The challenge of big data is how to use it to create something valuable to the user by this analysis helpful for people providing an accurate medical care and saving the life. To avoid the death rate, there should be a proper treatment provided to patient through posting in social networks with emergency communication management which helps related areas through request during disasters. Natural disasters are caused due to natural hazards and resistless unsafe communities that are irresponsible of adversities that are caused to them [1]. Human beings and animals dependently face the problems that are caused due to natural and human-made calamities and due to this loss of people, increase in economic losses, human suffering, enormous damages, and changes in environment. Predicting the natural calamities, losses on impacts of people and property are difficult with specific acceptance accuracy. Through the limited resources, it cannot forecast leads to take an adequate resources in advance due to change in climate, and it is difficult to predict the human life and damage caused by natural disasters. Bring up disaster management policies with combination of information technology by applying suitable levels and this equipment gives the huge potential in gathering the capabilities of disaster policies. Adding upon this, it also extends with latest technological resources that decrease the disaster risk [2]. Big data cooperates with researches to perform efficient results on vast amount on dataset. Big data is not only used for storage of data but subjected to accessing data, distribution of data with useful representation of data and analysis. This big data had come up with a consequential

T. Nirmala (✉) · S. A. Kumar · P. Rithvik Rao · P. Raviteja Reddy · T. Poojitha
MLR Institute of Technology, Dundigal, Hyderabad, India
e-mail: nirmala99teegala@gmail.com

V. Garcia Diaz and G. J. Rincón Aponte (eds.), *Confidential Computing*, Advanced Technologies and Societal Change,
https://doi.org/10.1007/978-981-19-3045-4_3

way of communication through exchange of intentional and output messages like a voice message or warning message is given to people before occurring the natural disaster by this people to start communicate with others reporting there situation to loved one for help [3]. This detailed analysis of vast information can be enhanced through big data accessed to researches on this outbreak in social media. Feasibility of project is analyzed in phase and proposal is put forth with very general plan on project by cost estimations while analyzing the system study on feasibility proposed a system that should handle this disaster analysis. It can fortify the recommended system is not freight to the organization. For viability survey, understanding the major components useful for the system that is essential. Applying machine learning techniques on dataset operations like data supervised unsupervised and reinforcement technique. ML provides a statistical tool for exploring and analyzing data, and this dataset is collection of natural and man-made calamities caused in various sectors like medical, fire accidents, and road accidents with this information prediction can be done and if any disaster occurs, it alerts the people before that takes place. Using this analysis, we can save people and economic losses. Unfortunately, if disaster occurs, instant steps to be taken to rescue their life by providing treatment through medical emergency calls, fire accident calls, and road accident calls and this can be known by social networks. With the use of Python, we develop a code and to get a consistent dataset by applying machine learning [4] techniques subset to data science where the machines are learned from dataset and predict the analysis. Through, data visualization techniques, vast amount of dataset can be handled through big data represented in graphs, pie charts, etc. It exert for both students, who so researches and hospitals, and people by getting this estimation records of areas where disasters occurs frequently will take this precautions with information that had got through past analysis by this improvising and developing the facilities in hospitals takes place.

Literature Survey

Analyzing historical data became easy due to advancement of analytical tools. Gathering data from social networking websites is a great challenge for today's data scientists. Many advancement and research has been conducted together streaming data [19].

On account of catastrophes or enormous scope crises, there were different calls to every organization mentioning for help, which expanded an opportunity to react and was hard to arrange [5].

Land-line and wireless calls were the best way to demand for help if there should be an occurrence of a crisis, and new advances were not incorporated Although every foundation put forth a critical attempt to furnish with the best support of the network, the absence of coordination and a brought together order was keeping FRIs from playing out their obligations viably [6].

During these visits, the group gathered related information about game-plan, innovations, and fruitful cases from the examination of PSAPs [7]. These cases were painstakingly examined by the Ecuadorian real factors, for example, socio-prudent conditions, the geographic association of the nation, previously existing establishments, utilization of new advancements, infiltration of cell phones and the Internet, utilization of informal organizations, and insights of mishaps and crises [8].

Existing System

In a developing country, facilities of each resource have to be updated with improving in information technology. The information which is contained in dataset is related to disasters but not the sectors in which it took place by this government cannot reduce the cases that mainspring disasters. Striding should be taken to avoid calamities. Predicatively, it displays only the total accuracy of processing the data by applying machine learning algorithms like Naive Bayes, KNN, etc. Analyzing dataset looks complicated because of their high volume [9] and complicated structure. Traditional database techniques have failed to handle these disasters dataset proficiently due to vast dataset.

In existing system, focusing on queries like how many manifestation based on fire/road accidents, etc. It gives exploratory details on why and how this emergency has occurred. Through population study and trained voice, it cannot load the record of accident that occurred.

Proposed System

Disaster analysis on government data consists of attributes in dataset with different disasters either natural or man-made calamities. This system causes the occurrence of every accident and has been recorded in dataset through particular type. It understands through reason that based on whether condition of population analysis and economical-social [10] happened. It gives us entire understanding of cause in events like zip code is related to mentioned accident which took place like madicent related to road accident. We could do it with help of machine learning techniques as part of data science. Some algorithms applied on dataset getting consistent data by cleaning, binning, etc., can performed based on linear regression and decision tree algorithm.

Dataset attributes of disaster analysis on management data with involvement of researches by providing resources from updates information technology [11]. Attributes that required to show prediction that it may or may not occur the disasters in particular area that people have to take care that who are suffering from natural calamities dataset attributes like latitude, longitude, description of emergency call, zip code, title, time stamp, township, and address.

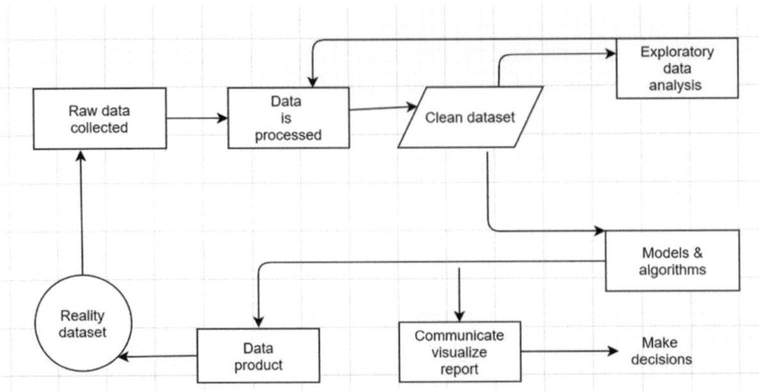

Fig. 3.1 Architecture

System Analysis

System Architecture

See Fig. 3.1.

System Requirements

Hardware requirements:

Processor	i3(min),i5
RAM	4 GB(min)
Hard Disk	1TB

Software requirements:

Operating System	window 7
Coding Language	Python
Editor	Jupiter
Command Prompt	Anaconda

Implementation

Implementing Algorithm

We are using efficient and nimble decision making algorithms on this dataset to get consistent by applying the modules on data provided through statistical tools for analyzing and exploring the information through classifying it in various sectors can be done through applying of ML techniques. Handling vast amount of data which can be updated daily basis report can be provided. Using that ML techniques, they classified in to three ways like supervised learning [11–13], unsupervised learning, and reinforcement learning as we use two algorithms for comparison of accuracy on analyzing the best accuracy with less precision on dataset which has represented by visualization techniques.

One of the algorithms in ML is decision tree, and it belonging to root of supervised learning where machine has trained with guide to get better analysis, i.e., model has instructed on dataset to start decision-making when it passed to testing data. It is used for collate the original input and expected output [14]. The motive of this algorithm is to create a training model based on analyzing the class or end variables by training decision rules it can also be solved by regression and classification too [15] (Fig. 3.2).

Another algorithm also used for implementation of accuracy rate on dataset related to disasters management. This techniques are used because researches are allowed to analyze the improving of information technology. So, putting every related news in social networks can be spread soon using these who shares there emergency situation to neighbor or loved one can help them by calling the emergencies with this we can save life and losses of economy. Linear regression model which defines the relationship between one dependent variable and one or more exploratory variables. As linear regression says, there will be a straight linear mark through dependent, i.e., scalar vector represented through x-axis but applying one are more independent

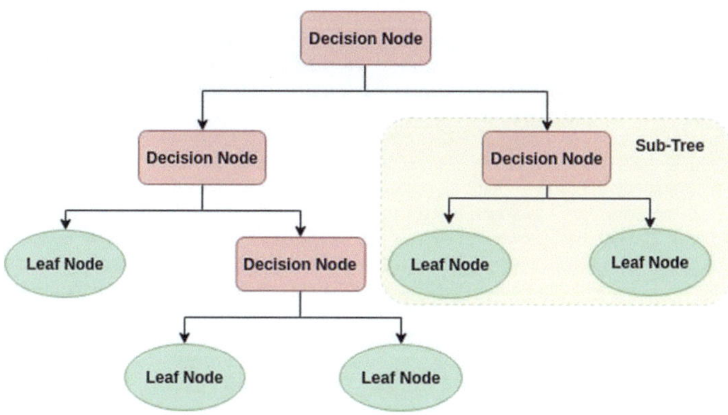

Fig. 3.2 Decision tree algorithm

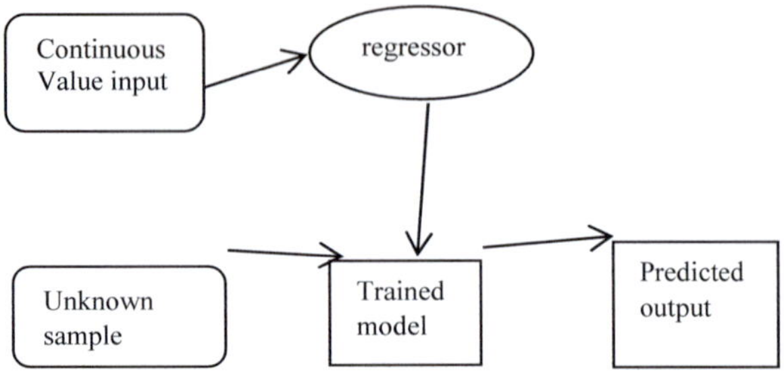

Fig. 3.3 Linear regression

variables also be exploratory variable may called as multivariate linear regression by this multiple correlated variables are predicted instead of single simple linear variant. If it has only one independent or exploratory variable called single linear regression (Fig. 3.3).

Evaluation and Results

Dataset

See Fig. 3.4.

1. Latitude
2. Longitude

lat	lng	desc	zip	title	timeStamp	twp	addr
40.2978759	-75.5813	REINDEER CT & DEAI	19525	EMS: BACK PAINS/INJURY	10-12-2015 17:10	NEW HANOVER	REINDEER CT & DEAD END
40.2580614	-75.2647	BRIAR PATH & WHIT	19446	EMS: DIABETIC EMERGEN	10-12-2015 17:29	HATFIELD TOWNSHIP	BRIAR PATH & WHITEMARSH LN
40.1211818	-75.352	HAWS AVE; NORRIS	19401	Fire: GAS-ODOR/LEAK	10-12-2015 14:39	NORRISTOWN	HAWS AVE
40.116153	-75.3435	AIRY ST & SWEDE ST;	19401	EMS: CARDIAC EMERGEN	10-12-2015 16:47	NORRISTOWN	AIRY ST & SWEDE ST
40.251492	-75.6033	CHERRYWOOD CT & DEAD END		EMS: DIZZINESS	10-12-2015 16:56	LOWER POTTSGROVE	CHERRYWOOD CT & DEAD END
40.2534732	-75.2832	CANNON AVE & W 9	19446	EMS: HEAD INJURY	10-12-2015 15:39	LANSDALE	CANNON AVE & W 9TH ST
40.1821111	-75.1278	LAUREL AVE & OAKD	19044	EMS: NAUSEA/VOMITING	10-12-2015 16:46	HORSHAM	LAUREL AVE & OAKDALE AVE
40.2172859	-75.4052	COLLEGEVILLE RD & I	19426	EMS: RESPIRATORY EMER	10-12-2015 16:17	SKIPPACK	COLLEGEVILLE RD & LYWISKI RD
40.2890267	-75.3996	MAIN ST & OLD SUM	19438	EMS: SYNCOPAL EPISODE	10-12-2015 16:51	LOWER SALFORD	MAIN ST & OLD SUMNEYTOWN PIKE
40.1023985	-75.2915	BLUEROUTE & RAMI	19462	Traffic: VEHICLE ACCIDEN	10-12-2015 17:35	PLYMOUTH	BLUEROUTE & RAMP I476 NB TO CHEM
40.2319898	-75.2519	RT202 PKWY & KNAPP RD; MOI		Traffic: VEHICLE ACCIDEN	10-12-2015 17:33	MONTGOMERY	RT202 PKWY & KNAPP RD
40.0841613	-75.3084	BROOK RD & COLWE	19428	Traffic: VEHICLE ACCIDEN	10-12-2015 16:32	PLYMOUTH	BROOK RD & COLWELL LN
40.1741312	-75.0985	BYBERRY AVE & S W	19040	Traffic: VEHICLE ACCIDEN	10-12-2015 17:15	UPPER MORELAND	BYBERRY AVE & S WARMINSTER RD
40.062974	-75.1359	OLD YORK RD & VAL	19027	Traffic: VEHICLE ACCIDEN	10-12-2015 17:12	CHELTENHAM	OLD YORK RD & VALLEY RD
40.0972222	-75.3762	SCHUYLKILL EXPY & CROTON RI		Traffic: VEHICLE ACCIDEN	10-12-2015 17:09	UPPER MERION	SCHUYLKILL EXPY & CROTON RD UNDEI
40.2237777	-75.2354	STUMP RD & WITCH	18936	Traffic: VEHICLE ACCIDEN	10-12-2015 17:31	MONTGOMERY	STUMP RD & WITCHWOOD DR
40.2432578	-75.2866	SUSQUEHANNA AVE	19446	EMS: RESPIRATORY EMER	10-12-2015 17:42	LANSDALE	SUSQUEHANNA AVE & W MAIN ST
40.3121807	-75.5743	CHARLOTTE ST & MII	19525	EMS: DIZZINESS	10-12-2015 17:41	NEW HANOVER	CHARLOTTE ST & MILES RD
40.114239	-75.3385	PENN ST & ARCH ST;	19401	EMS: VEHICLE ACCIDENT	10-12-2015 17:43	NORRISTOWN	PENN ST & ARCH ST
40.2093369	-75.1353	COUNTY LINE RD & V	18974	Traffic: DISABLED VEHICLI	10-12-2015 17:45	HORSHAM	COUNTY LINE RD & WILLOW DR
40.114239	-75.3385	PENN ST & ARCH ST;	19401	Traffic: VEHICLE ACCIDEN	10-12-2015 17:43	NORRISTOWN	PENN ST & ARCH ST

Fig. 3.4 Dataset

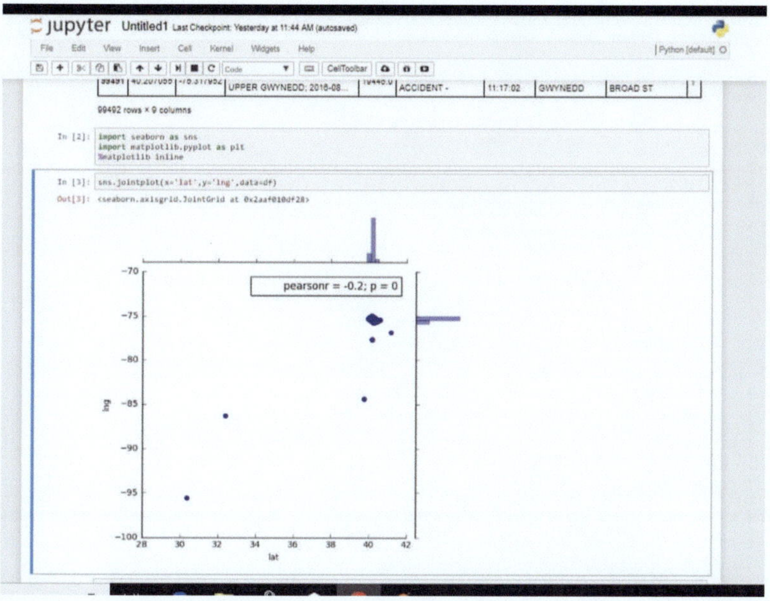

Fig. 3.5 Joint plot for visualizing relation between latitude and longitude

3. Description
4. Zip code
5. Title
6. Time Stamp
7. Township
8. Address

Result of the Analysis

See Figs. 3.5, 3.6, 3.7, 3.8, and 3.9

Conclusion

In our project, mainly useful for both hospital to improve their medical treatment to patients and people on taking early precautions before occurrence of natural or man-made disasters. By displaying analysis through visualization techniques, it can explore by applying different techniques by creating data frames, identifying it, expanding and representing through bar graphs finally we want visualization. In this dataset, we can identify the EMS linked phones by this we can say that mostly

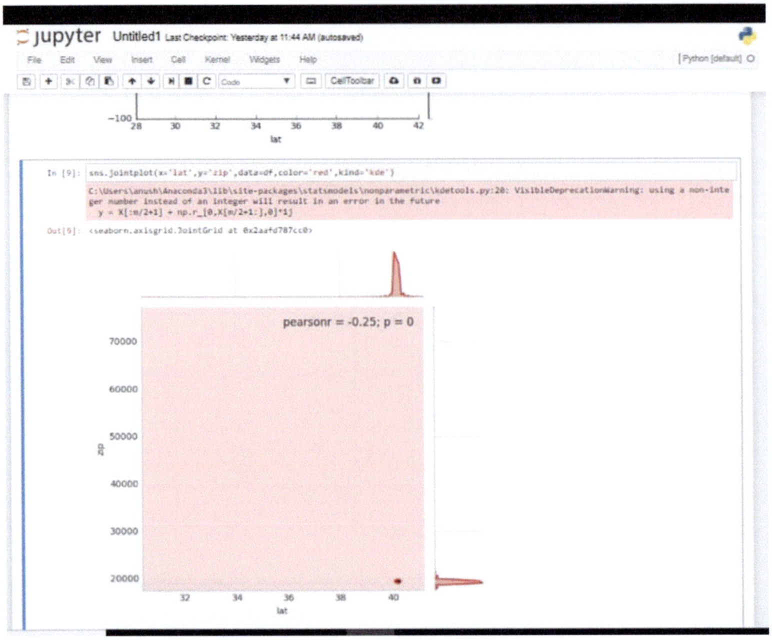

Fig. 3.6 Joint plot for visualizing relation between latitude and zip code

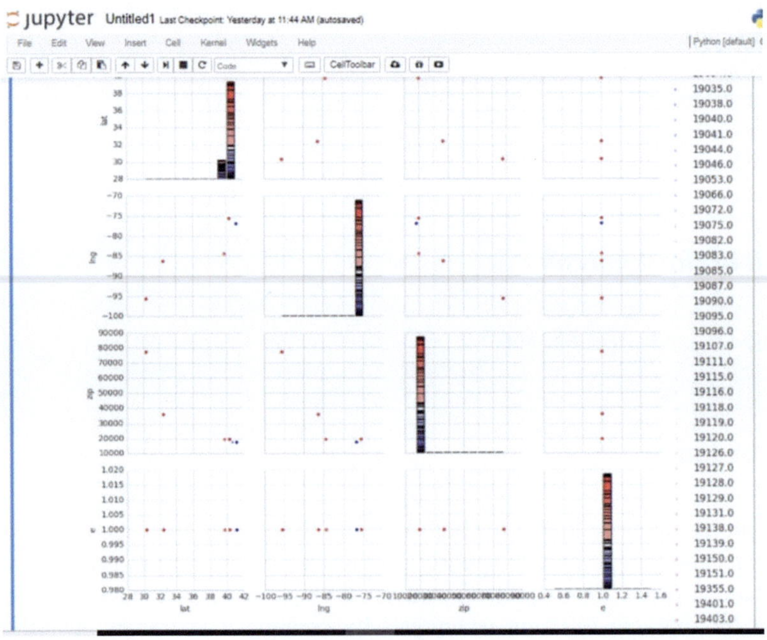

Fig. 3.7 Scatter plot

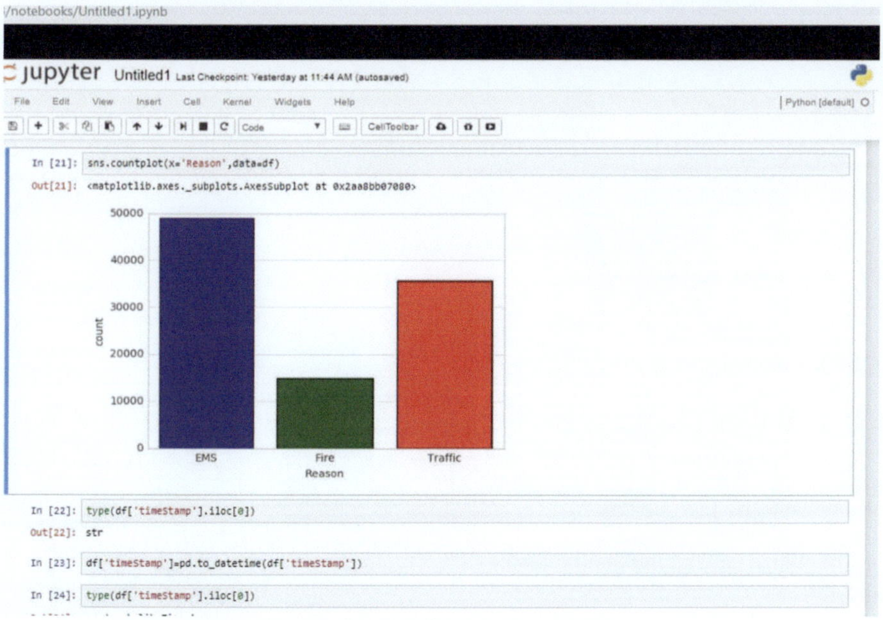

Fig. 3.8 Result graph showing reason for disasters

disasters occurred area. We found some cases in dataset took place on 2015–16 due to natural calamities you can browse the information on given years. Linear regression is mostly important for this disaster analysis.

Future Scopes

This project which involves development of data prediction and machine learning techniques. Future enhancement of this to elaborate it by making more advanced through adding some graphical user interfaces by adding new libraries are packages in Python or *R*. With this implementation, we can attract the users by ease understanding of graphs. It can also been draw much concentrated conclusions by adding results that we got from computer vision and usage of required hardware by opened up sources for development efficiency through ML-based implementations by using graded analysis.

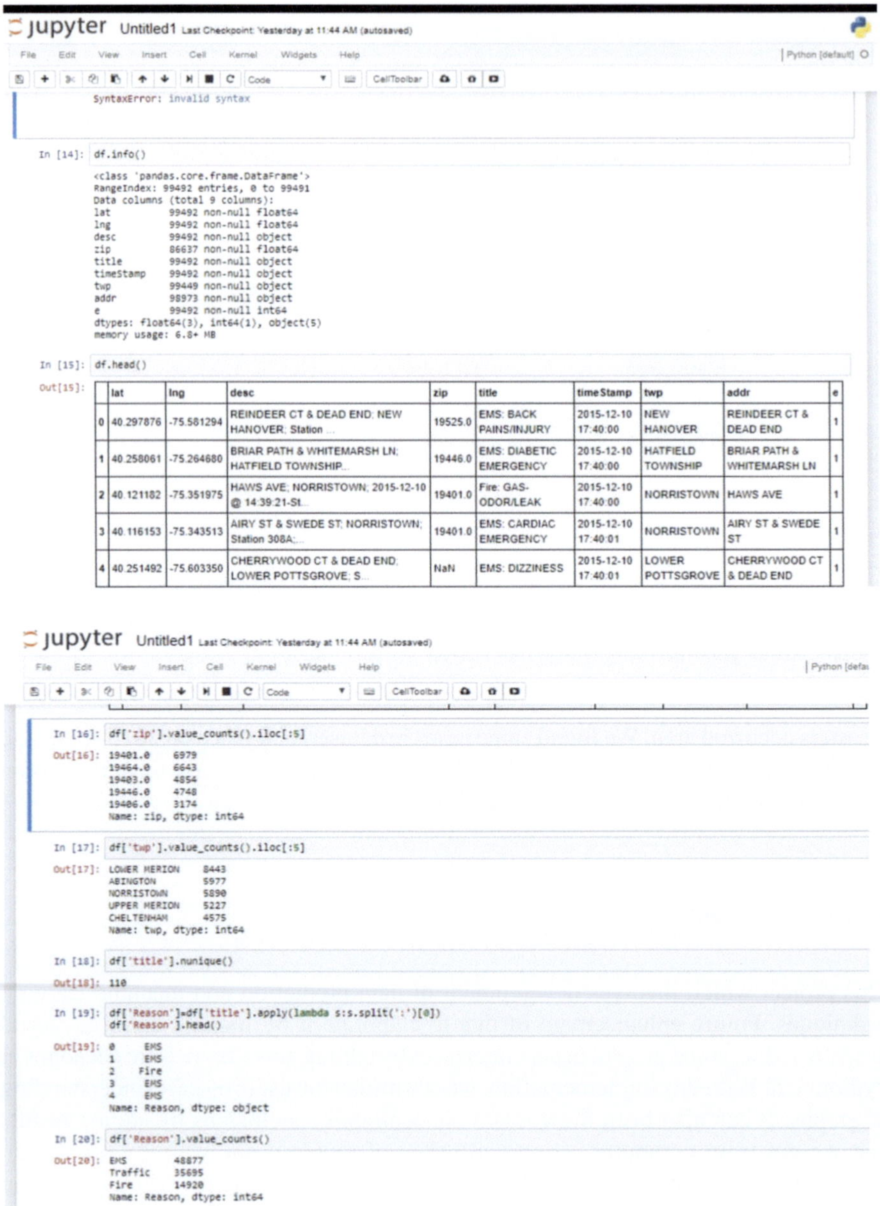

Fig. 3.9 Attributes of dataset

References

1. Sai Prasad, K., Pasupathy, S.: Real-time data streaming using apache spark on fully configured hadoop cluster. J. Mech. Cont. Math. Sci. **13**(5), 164–176 (2018)
2. -1–1 origin amp history—national emergency number association (2017)
3. Ahmed, M., Ahmed, R., Thakuria, A.J., Laskar, R.H.: Eye center guided constrained local model for landmark localization in facial image. In: 2019 9th Annual Industrial Automation and Electromechanical Engineering Conference (IEMECON). IEEE (2019)
4. Arnold, T., Tilton, L.: Natural Language Processing, pp. 131–155. Springer International Publishing, Cham (2015)
5. Sowmya, G., Divya Jyothi, G., Shirisha, N., Navya, K.: Active learning strategies in engineering education. J. Adv. Res. Dyn. Control Syst. Special Issue (03), 1253–1258 (2018)
6. Prasad, K.S., Reddy, N.C.S., Puneeth, B.N.: A framework for diagnosing kidney disease in diabetes patients using classification algorithms. SN Comput. Sci. **1**, 101 (2020)
7. Divya Jyothi, G., Navya, K.: Design and implementation of a store management system. In: 2017 International Conference on Intelligent Sustainable Systems (ICISS), pp. 1149–1151 (2017)
8. Lakshmi, B., Madhuravani, B., Veda Vidya, B., Sowjanya, C.: SeSPHR: a methodology for secure sharing of personal health records in the cloud. Int J IJRTE **2**(6), Print ISSN: 2395-1990, Online ISSN : 2394-4099; 30 2019, Page No.690–694.
9. Dean, J., Ghemawat, S.: 'MapReduce: Simplified data processing on large clusters.' Commun. ACM **51**(1), 107–113 (2008)
10. Bhavani, M., Narayana, V.A., Sreevani, G.: A novel approach for detecting near-duplicate web documents by considering images, text, size of the document and domain. In: Lecture Notes in Electrical Engineering, vol. 398, pp. 1355–1366 (2021)
11. De, S., Zhou, Y., Abad, I.L., Moessner, K.: Cyber–physical–social frameworks for urban big data systems: a survey. Appl. Sci. **7**(10), 1017 (2017)
12. LeCun, Y., Bengio, Y., Hinton, G.: Deep learning. Nature **521**, 436–444 (2015)
13. Merugu, S., Tiwari, A., Sharma, S.K.: Spatial–spectral image classification with edge preserving method. J. Indian Soc. Remote Sens. (2021). ISSN 0255-660X. https://doi.org/10.1007/s12524-020-01265-7
14. Naik, M.V., Ansari, M.D., Gunjan, V.K., Surya Narayana, G.: An approach for morphological analyzer rules for dravidian Telugu language. In: Lecture Notes in Electrical Engineering, vol. 398, 1385–1392 (2021)
15. Gogineni, S., Pimpalshende, A., Goddumarri, S.: Eye disease detection using yolo and ensembled googlenet. In: Lecture Notes on Data Engineering and Communications Technologies, vol. 53, 465–482 (2021)

Chapter 4
EDF: An Enhancement of Droid Fusion Framework for Mitigation of Multi-class Malware

A. Sangeetha and P. Upendar

Introduction

The usage of mobile devices is increasing vastly in the present day. The mobile applications are spread in different industry fields such as health care, education, business and etc. The many mobile applications are implemented on the android operating system, it shared 86% of market demand [1, 2]. So high mobile applications depend on the android. In Android applications, maintainenance of security and privacy is the main concern. Because many malware mobile applications are targeted on android mobile devices. The Android malware applications attack mobile devices and compromise mobile nodes to break the security and privacy data sources. The compromise nodes also allow unauthorized access and perform malware activities [3].

The advancement in mobile application development such as high pixel cameras and online service with the linkage of GPS services leads to an increase in the malicious attacks. The academic and industry researchers implemented many machine learning algorithms for attention security issues. They mostly focus on application-level verification such as security, license, and intrusion. The malware applications have different type of categories, which are secretly embedded in mobile applications and perform harmful activities.

In recent research to classify the malware introduced a Droid Fusion framework based on machine learning algorithms with the features of the multi-level architecture. The Droid Fusion worked on two different levels. At the lower level the android

A. Sangeetha (✉) · P. Upendar
Department of Computer Science and Engineering, MLR Institute of Technology, Hyderabad, India
e-mail: allamsangeetha@gmail.com

P. Upendar
e-mail: upendar.para@gmail.com

V. Garcia Diaz and G. J. Rincón Aponte (eds.), *Confidential Computing*,
Advanced Technologies and Societal Change,
https://doi.org/10.1007/978-981-19-3045-4_4

malware is detected by the training number of base classifiers and at the higher level set of ranking-based algorithms are used. This framework improved accuracy in the classification of android malware. However, Droid Fusion framework does not reach the expectation of the real-time industry and also it doesn't support the multi-class classification of android malware [4]. Achieving high performance and improve the mitigation of malware in mobile applications are essential. The enhancement of the Droid Fusion framework is the need for an emergency in mobile applications [5, 6].

In this research work, we proposed an EDF framework. This framework overcomes the limitations of Droid Fusion and improve performance [7]. This framework utilizes machine learning and ranking-based algorithms in the mitigation of android malware applications [8]. This framework solves the problems with multi-class malware. The empirical results of the EDF framework showed better improvement with the comparison of benchmark frameworks such as Droid Fusion, MultiScheme, J48, REPTree and etc.

The remainder of the paper is organized as follows: Section "Literature Survey" given the discussion of the related works. We present the methodology of the proposed framework in Section "Proposed Methodology". Section "Results" shows the comparison of experimental results. Finally, in Section "Conclusion", we conclude the research work.

Literature Survey

Many research frameworks are introduced for detection of malwares in android mobile applications. In that some of research frameworks used static techniques and some frame works utilize dynamic techniques or both techniques. All the machine learning-based frameworks addressed the limitations of security problems in android mobile applications. The main malware activities in android mobile applications are privacy maintenance and data leakage.

Wu et al. [9], introduced DroidMat framework to classify the android mobile applications are benign or malware. The DroidMat built based on the *k*-Means and *k*-Nearest Neighbor algorithms. This framework also used static features to installation of mobile applications.

Yerima et al. [10], introduced high malware detection rate framework called DAPASA which is developed using the machine learning algorithms such as Random Forest, Decision Tree, *k*-NN and PART. The DAPASA framework utilizes the sensitive sub graphs to detect android malwares in piggybacked. This framework performs the better performance but has the limitations of static features.

Fan et al. [11], introduced android malware detection by utilizing the parallel machine learning algorithms. In this paper they performed comparison classifier fusion methods using J48, Naïve Bayes, PART, RIDOR. The research work explored that the fusion models given better performance when compare with single classifiers.

Wang et al. [12] introduced ensemble classifiers, these classifiers utilized static features other classifiers such as SVM, *k*-NN, Naïve Bayes, CART and Random

Forest. The empirical results showed that the ensemble classifiers are more secure with the comparison of individual classifiers.

Idress et al. [13], proposed a novel android malware detection framework called as PIndroid. This framework utilizes the ensemble learning methods. To improve the performance this framework utilizes the machine learning models and classifier fusion.

Proposed Methodology

Problem Statement

Droid Fusion framework is designed with the combination of special static features and classifiers. In android malware detection the multi-level architecture framework performed improved accuracy. The main aim of the Droid Fusion framework is binary classification. This classification fusion had high malware detection rate, but it's unable to the solve the problems with multi-class malwares. This research paper proposed EDF framework to overcome the limitations of Droid Fusion framework.

Enhanced Droid Fusion

The machine learning algorithms are essential in malware detection of android mobile applications. The binary classification [14] fusion mitigates the malwares in the two classes only. The proposed EDF framework mitigates the multi-class malwares. The EDF framework implementation follows the below steps (Fig. 4.1).

Feature Selection

The EDF framework takes the account of loaded applications into the framework. On each and every loaded application the machine learning algorithm applied at low level. Here calculate the multiple features of android mobile application. Then the process goes to high level. In high level ranking-based algorithm is applied for basic prediction of android applications.

EDF Classification

In the phase of feature selection base classifiers applied on the android mobile applications based on the multiple features using the ranking. The EDF frameworks are applied on resultant mobile application of the feature selection phase. In this phase

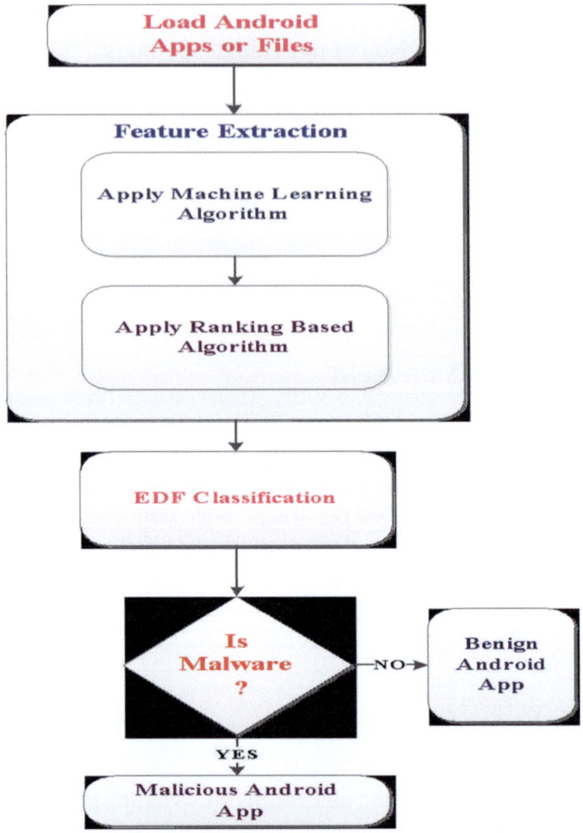

Fig. 4.1 Implementation flow of EDF framework

the Enhanced Droid Fusion classifier is applicable for detection or mitigation of android mobile applications [15]. The EDF framework classifies and mitigates the multi-class malwares such as spyware, rootkit, ransomware, etc.

Algorithm

Algorithm Name: Enhancement Droid Fusion Algorithm

(continued)

(continued)

Input: Android Mobile Apps(AMP$_n$)
Output: Detected Android Malware Apps
Load Android Mobile Apps
Apply MLA on AMPn
Extract Features
Apply RBA on AMPn
Find Rank
Apply EDF on Resultant Apps
IF (Check Multi-class Malware) {
Detected Malwares
} else {
No Malware Found
}
End

Results

In this section, we discuss about the empirical results of the proposed EDF framework. The proposed EDF fusion classifier results compare with some of the benchmark classifiers such as J48, REPTree, MultiScheme, Droid Fusion.

Fig. 4.2 showed that the proposed EDF framework achieves better performance with the comparisons of existing fusion frameworks. The empirical results showed that the proposed framework achieved high weighted feature measurement score i.e.

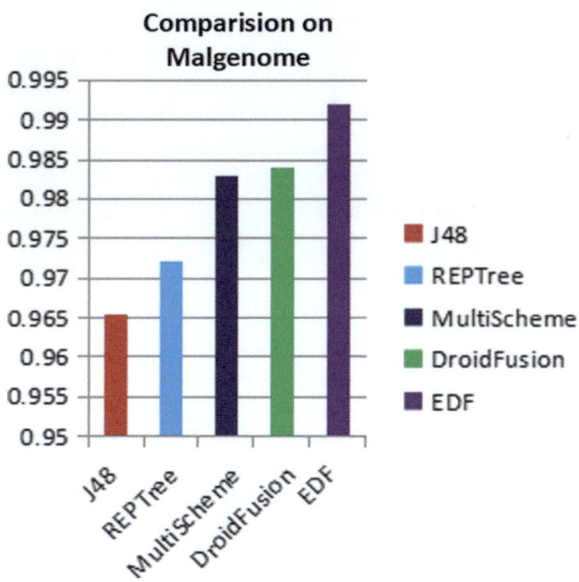

Fig. 4.2 Comparison of classifiers on Malgenome dataset

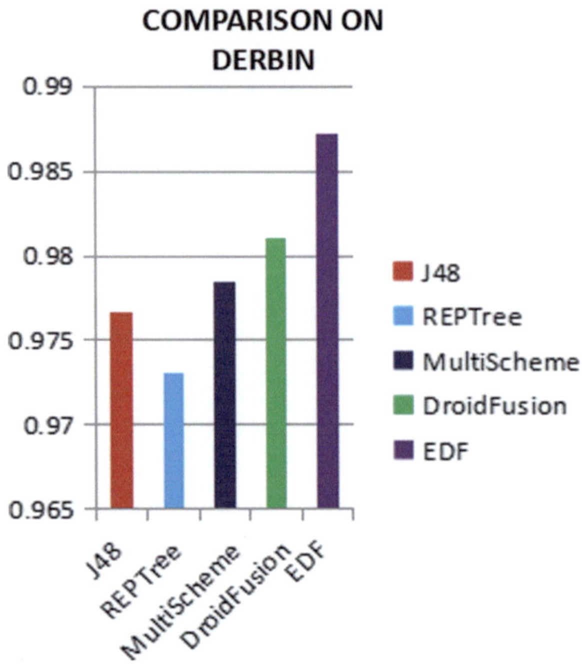

Fig. 4.3 Comparison of classifiers on DERBIN dataset

WFM = 0.99, with the comparisons of other Droid Fusions [16]. The comparison is performed on the Malgenome dataset.

As showed in Fig. 4.3 that the proposed EDF framework achieves better performance with the comparisons of existing fusion frameworks. The empirical results showed that the proposed framework achieved high weighted feature measurement score i.e. WFM = 0.977, with the comparisons of other Classifiers. The comparison is performed on the DERBIN dataset.

In Fig. 4.4 showed that the proposed EDF framework achieves better performance with the comparisons of existing fusion frameworks.

The empirical results showed that the proposed framework achieved high weighted feature measurement score i.e. WFM = 0.9822, with the comparisons of other Droid Fusions. The comparison is performed on the MCAFEE dataset.

Conclusion

This research work concludes that the recently a Droid Fusion framework is introduced for malware detection in android mobile applications. Even though the Droid Fusion framework has given better classification results this framework is applicable for static features and not for multi-class malware mitigation. To address the limitations of the Droid Fusion framework, we proposed the enhancement of the Droid

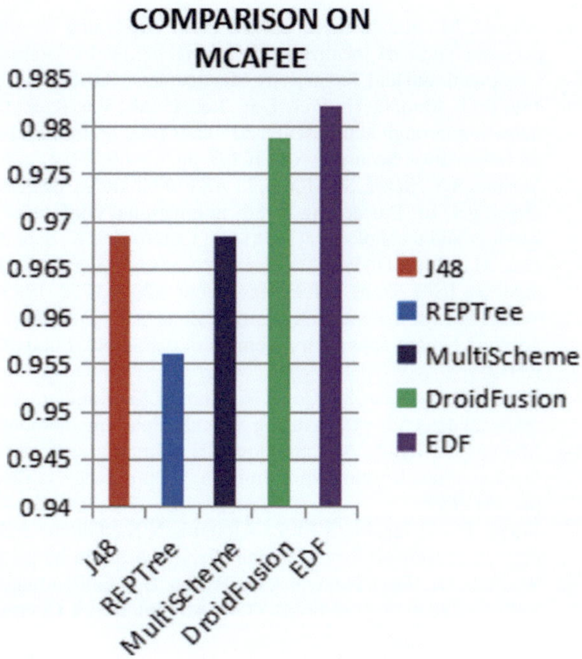

Fig. 4.4 Comparison of classifiers on MCAFEE dataset

Fusion (EDF) framework. The proposed framework mitigates the malware attacks effectively. The results of the proposed EDF framework achieved high performance with the comparison of the benchmark of android malware detection frameworks.

References

1. McAfee Labs Threat Predictions Report, McAfee Labs, Santa Clara, CA, USA, Mar 2016
2. Arp, D., Spreitzenbarth, M., Hubner, M., Gascon, H., Rieck, K.: Drebin: efficient and explainable detection of Android malware in your pocket. In: Proceeding of 20th Annual Network Distribution System Security Symposium (NDSS), San Diego, CA, USA, Feb 2014, pp. 1–15
3. Yerima, S.Y., Sezer, S.: Droid Fusion: a novel multilevel classifier fusion approach for android malware detection. IEEE Trans Cybern 1–14 (2019)
4. Narayana, V.A., Premchand, P., Govardhan, A.: A novel and efficient approach for near duplicate page detection in web crawling. In: 2009 IEEE International Advance Computing Conference, IACC 2009, 2009
5. Yerima, S.Y., Sezer, S., Muttik, I.: Android malware detection: an eigenspace analysis approach. In: Proceedings of Science and Information Conference (SAI), London, UK, Jul 2015, pp. 1236–1242
6. Choudhary, S.R., Gorla, A., Orso, A.: Automated test input generation for Android: Are we there yet? In: Proceedings of 30th IEEE/ACM International Conference on Automated Software Engineering (ASE), Nov 2015, pp. 429–440
7. Ho, T.K.: Random decision forests. In: Proceeding of 3rd International Conference on Document Analysis and Recognition, pp. 278–282, 1995

8. Ahmed, M., Karsh, R.K., Laskar, R.H.: Analyzing the effect of eye center localization on accurate landmark localization in a facial image. In: International Conference on Automation, Computational and Technology Management, ICACTM 2019, London. IEEE
9. Wu, D.-J., Mao, C.-H., Wei, T.-E., Lee, H.-M., Wu, K.-P.: (2012) DroidMat:Android malware detection through manifest and API calls tracing. In: Proceeding of 7th Asia Joint Conference on Information Security (Asia JCIS), pp. 62–69, 2012
10. Yerima, S.Y., Sezer, S., Muttik, I.: Android malware detection using parallel machine learning classifiers. In: Proceeding of 8th international conference on next generation mobile apps, services and technologies (NGMAST), Oxford, UK, Sept 2014, pp. 37–42
11. Fan, M., et al.: DAPASA: detecting Android piggybacked apps through sensitive subgraph analysis. IEEE Trans. Inf. Forens. Secur. **12**(8), 1772–1785 (2017)
12. Wang, W., Li, Y., Wang, X., Liu, J., Zhang, X.: Detecting Android malicious apps and categorizing benign apps with ensemble of classifiers. Future Gener. Comput. Syst. **78**, 987–994 (2017)
13. Idrees, F., Rajarajan, M., Conti, M., Chen, T.M., Rahulamathavan, Y.: PIndroid: a novel Android malware detection system using ensemble learning methods. Comput. Secur. **68**, 36–46 (2017)
14. Merugu, S., Reddy, M.C.S., Goyal, E., Piplani, L.: Text message classification using supervised machine learning algorithms. In: Lecture Notes in Electrical Engineering 2019, vol. 500, pp. 141–150
15. Sihag, P., Al-Janabi, A.M.S., Alomari, N.K., Ghani, A.A., Nain, S.S.: Evaluation of tree regression analysis for estimation of river basin discharge. Model Earth Syst Environ 2021
16. Saikiran, G., Surya Narayana, G., Porika, D., Vinit Kumar, G.: Clinical skin disease detection and classification: ensembled VGG. Adv Intell Syst Comput, vol. 1245, pp. 827–847 (2021)

Chapter 5
Early Prediction of Chronic Kidney Disease Using Predictive Analytics

B. Madhuravani, R. Krishnasrija, and Divya Priya Degala

Introduction

The objective of this proposed work is to analyze different parameters that can lead to the condition of chronic kidney disease and understand the relation between them to draw a conclusion about the chronic kidney disease. In this study, we inspect the capacity of a few AI techniques based on certain algorithms for early forecast of the disease. Chronic kidney disease is a long lasting condition that incited by reduced function of the kidney.

Early detection and legitimate medicines can stop or moderate the movement of this constant ailment to end stage, where transplantation of damaged and infected kidneys or using artificial methods like dialysis are the only ways to spare patient's life. In most cases, it happens that the patients get to know about their disease only after noticing symptoms like fever, water retention, etc. By this time, the disease might have infected and damaged one or both the kidneys leading the patient to go through most difficult procedures like dialysis or kidney transplantation. So, to prevent such cases, we would like to build a reliable mechanism to predict the disease even before they notice the symptoms and before the kidneys get damaged. We are supporting our technique by the utilization of prescient investigation, in which we look at the relationship in the middle of information parameters just as with the objective class characteristic. Prescient examination empowers us to present the ideal subset of parameters to take care of AI to construct a lot of prescient models.

B. Madhuravani (✉) · R. Krishnasrija · D. P. Degala
Department of Computer Science and Engineering, MLR Institute of Technology, Dundigal, Hyderabad, Telangana, India
e-mail: peddi.madhuravi@gmail.com

V. Garcia Diaz and G. J. Rincón Aponte (eds.), *Confidential Computing*,
Advanced Technologies and Societal Change,
https://doi.org/10.1007/978-981-19-3045-4_5

Literature Survey

During the literature survey [1–6], we surveyed different sources and pulled out information about the mechanisms used currently for predicting the chronic kidney disease.

In 2016, Padmanaban and Parthiban [7] worked to predict the disease for patients suffering with diabetes using machine learning methods. The authors conducted tests on the dataset utilizing the decision tree along with Naïve Bayes methods classify. They came to a conclusion that that the decision tree algorithm is more optimal than the Naïve Bayes.

Salekin and Stankovic [8–10] have evaluated three classification techniques. By utilizing a method called wrapper, a component decrease examination was done to discover the properties that distinguish this sickness with high exactness. By considering few main features, they can anticipate the disease.

Yildirim [11, 12] inspected the impact of testing calculations in anticipating interminable kidney illness. The examination was finished by contrasting the impact of the three inspecting calculations. The examination indicated that inspecting calculations could improve the grouping calculation execution, and the resample strategy has a higher exactness among the testing calculations [13].

A. *Existing System*

From the literature survey [14], it is clear that classification-based algorithms have a great ability to categorize and predict the disease in a person. The present systems based on models built which use different algorithms like the support vector classifier, multilayer perceptron, lda, guas, etc. have been relied upon and are giving certain results which are accurate until certain percentage. In the proposed work, we would like to put forward more appropriate algorithm which would give more accuracy compared to the mentioned existing ones.

B. *Proposed System*

In the proposed system, we have two phases. In the first phase, we have built a model based on an algorithm called decision tree classifier. As an output, a decision tree is generated with yes and no as leaf nodes indicating that the class label chronic kidney disease is predicted in the case or not.

To build a decision tree, information gain is required to take a decision on the split of which feature are we going to build the tree; at every step, we must choose the split that generates the purest form of daughter nodes. We have taken gini index and entropy as information gain sources in order to build the decision tree. So, two decision trees are obtained with gini index and entropy as information gain sources.

The second phase is an evaluation phase, where we compare our decision tree classifier algorithm-based model with other existing algorithms like svc, log reg, etc. to strengthen the system and model we have developed in our first phase (Fig. 5.1).

The first step involved in the model is taking a dataset which refers to the gathering of data. Later, the data is preprocessed to make the data set free from invalid, incorrect

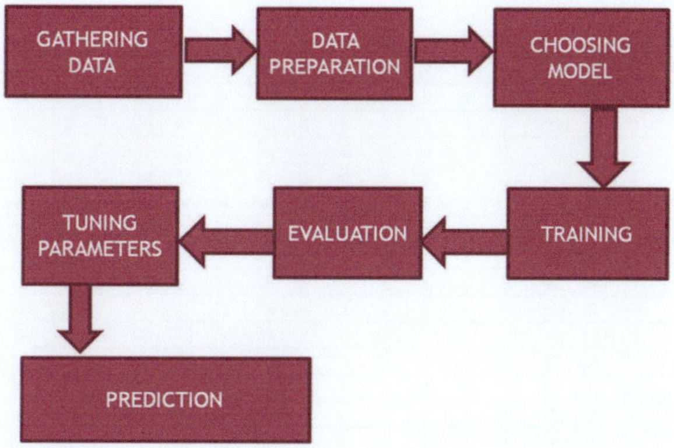

Fig. 5.1 System architecture

and incomplete parts. Then, the suitable algorithm is chosen, which is called decision tree classifier and a model is built on it. The dataset is divided into training and test data and then the training is done based on the data. After that, the model built is used to test the remaining data. This is called evaluation, and after that, parameters are tuned accordingly. Then, the model is used to predict new data values. So, this is the architecture and function of the system (Fig. 5.2).

The dataset is taken from UCI open source with 24 attributes and a class label indicating yes or no for chronic kidney disease existence (Fig. 5.3).

Evaluation and Results

The decision trees shown are obtained based on information gain sources entropy and gini index. The leaf nodes give the class label yes or no for the disease prediction. These are the results of the first phase of the proposed work (Figs. 5.4 and 5.5).

The feature importance graph depicts the importance of each attribute on which the chronic kidney disease depends and which attributes must be focused upon to predict the disease (Figs. 5.6 and 5.7).

The most crucial part of second phase is the comparison of decision tree classifier with the other algorithms and observing their accuracies based on training and test set scores (Fig. 5.8).

So, after observing all the training and test set scores, it is clear that the decision tree classifier has got the highest score accuracies. Now, it can be said that the decision tree classifier and the model we built out of it outweighs models built on the other algorithms. So, you can see the result printing the decision tree classifier as the best out of them.

age	bp	sg	al	su	rbc	pc	pcc	ba	bgr	bu	sc	sod	pot	hemo	pcv	wbcc	rbcc	htn	dm	cad	appet	pe	ane	class
48	80	1.02	1	0	0	0	0	0	121	36	1.2	0	0	15.4	44	7800	5.2	1	1	0	1	0	0	1
7	50	1.02	4	0	0	0	0	0	0	18	0.8	0	0	11.3	38	6000	0	0	0	0	1	0	0	1
62	80	1.01	2	3	0	0	0	0	423	53	1.8	0	0	9.6	31	7500	0	0	1	0	0	0	1	1
48	70	1.005	4	0	0	1	1	0	117	56	3.8	111	2.5	11.2	32	6700	3.9	1	0	0	0	1	1	1
51	80	1.01	2	0	0	0	0	0	106	26	1.4	0	0	11.6	35	7300	4.6	0	0	0	1	0	0	1
60	90	1.015	3	0	0	0	0	0	74	25	1.1	142	3.2	12.2	39	7800	4.4	1	1	0	1	1	0	1
68	70	1.01	0	0	0	0	0	0	100	54	24	104	4	12.4	36	0	0	0	0	0	1	0	0	1
24	0	1.015	2	4	0	1	0	0	410	31	1.1	0	0	12.4	44	6900	5	0	1	0	1	1	0	1
52	100	1.015	3	0	0	1	1	0	138	60	1.9	0	0	10.8	33	9600	4	1	1	0	1	0	1	1
53	90	1.02	2	0	1	1	1	0	70	107	7.2	114	3.7	9.5	29	12100	3.7	1	1	0	0	0	1	1
50	60	1.01	2	4	0	1	1	0	490	55	4	0	0	9.4	28	0	0	1	1	0	1	0	1	1
63	70	1.01	3	0	1	1	1	0	380	60	2.7	131	4.2	10.8	32	4500	3.8	1	1	0	0	1	0	1
68	70	1.015	3	1	0	0	1	0	208	72	2.1	138	5.8	9.7	28	12200	3.4	1	1	1	0	1	0	1
68	70	0	0	0	0	0	0	0	98	86	4.6	135	3.4	9.8	0	0	0	1	1	1	0	1	0	1
68	80	1.01	3	2	0	1	1	1	157	90	4.1	130	6.4	5.6	16	11000	2.6	1	1	1	0	1	0	1
40	80	1.015	3	0	0	0	0	0	76	162	9.6	141	4.9	7.6	24	3800	2.8	1	0	0	1	0	1	1
47	70	1.015	2	0	0	0	0	0	99	46	2.2	138	4.1	12.6	0	0	0	0	0	0	1	0	0	1
47	80	0	0	0	0	0	0	0	114	87	5.2	139	3.7	12.1	0	0	0	1	0	0	0	0	0	1
60	100	1.025	0	3	0	0	0	0	263	27	1.3	135	4.3	12.7	37	11400	4.3	1	1	1	1	0	0	1
62	60	1.015	1	0	0	1	1	0	100	31	1.6	0	0	10.3	30	5300	3.7	1	0	1	1	0	0	1
61	80	1.015	2	0	1	1	0	0	173	148	3.9	135	5.2	7.7	24	9200	3.2	1	1	1	0	1	1	1
60	90	0	0	0	0	0	0	0	0	180	76	4.5	0	10.9	32	6200	3.6	1	1	1	1	0	0	1
48	80	1.025	4	0	0	1	0	0	95	163	7.7	136	3.8	9.8	32	6900	3.4	1	0	0	1	0	1	1
21	70	1.01	0	0	0	0	0	0	0	0	0	0	0	0	0	0	0	0	0	0	0	0	1	1
42	100	1.015	4	0	0	1	0	1	0	50	1.4	129	4	11.1	39	8300	4.6	1	0	0	0	0	0	1
61	60	1.025	0	0	0	0	0	0	108	75	1.9	141	5.2	9.9	29	8400	3.7	1	1	0	1	0	1	1
75	80	1.015	0	0	0	0	0	0	156	45	2.4	140	3.4	11.6	35	10300	4	1	1	0	0	0	0	1
69	70	1.01	3	4	0	1	0	0	264	87	2.7	130	4	12.5	37	9600	4.1	1	1	1	1	1	0	1
75	70	0	1	3	0	0	0	0	123	31	1.4	0	0	0	0	0	0	0	1	0	1	0	0	1
68	70	1.005	1	0	1	1	1	0	0	28	1.4	0	0	12.9	38	0	0	0	0	1	1	0	0	1

Fig. 5.2 Sample dataset

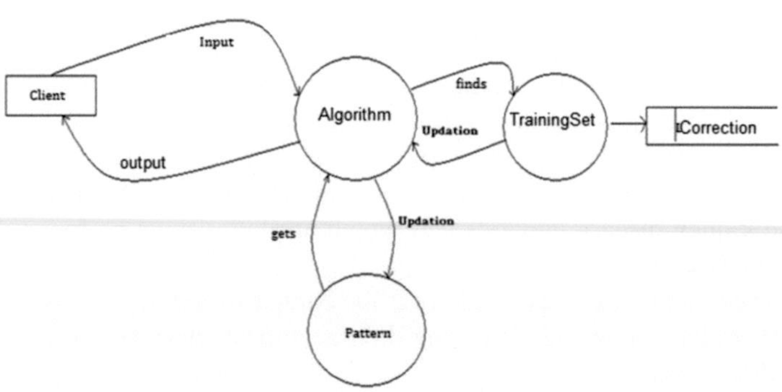

Fig. 5.3 Data flow diagram

Conclusion

Results illustrate that machine learning using the decision tree classifier as the most suitable and appropriate among the other algorithms. We would like to use the model

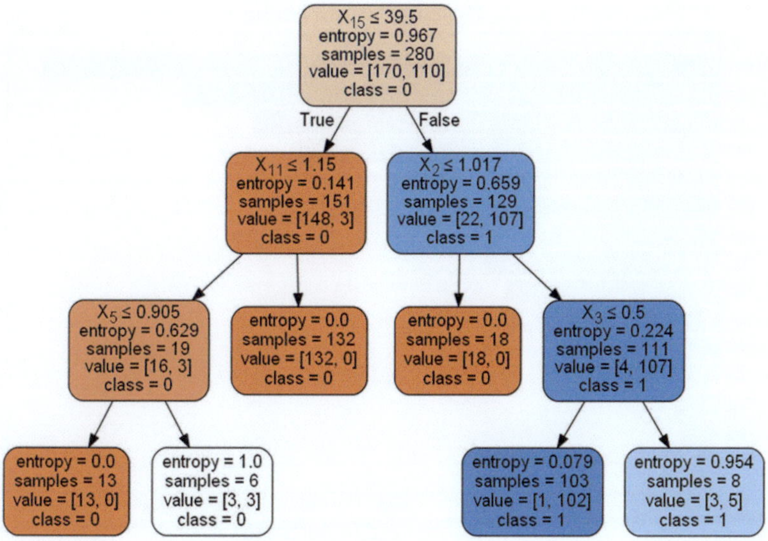

Fig. 5.4 Decision tree based on entropy

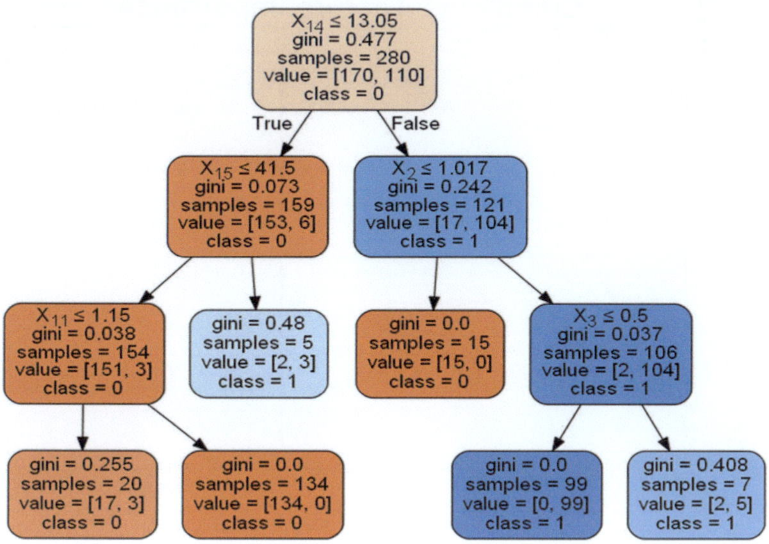

Fig. 5.5 Decision tree based on gini index

in nephrology specialized departments. Along with this, we would like to make it available for physicians in general checkups before the symptoms are visible.

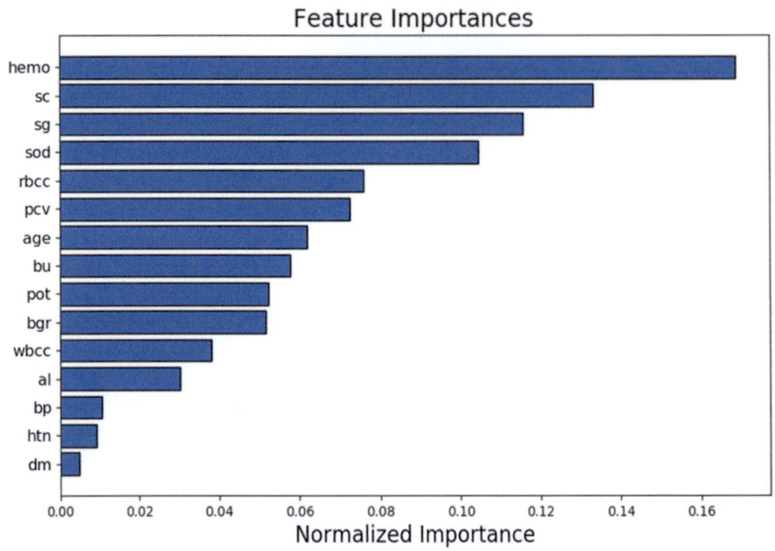

Fig. 5.6 Feature importance graph

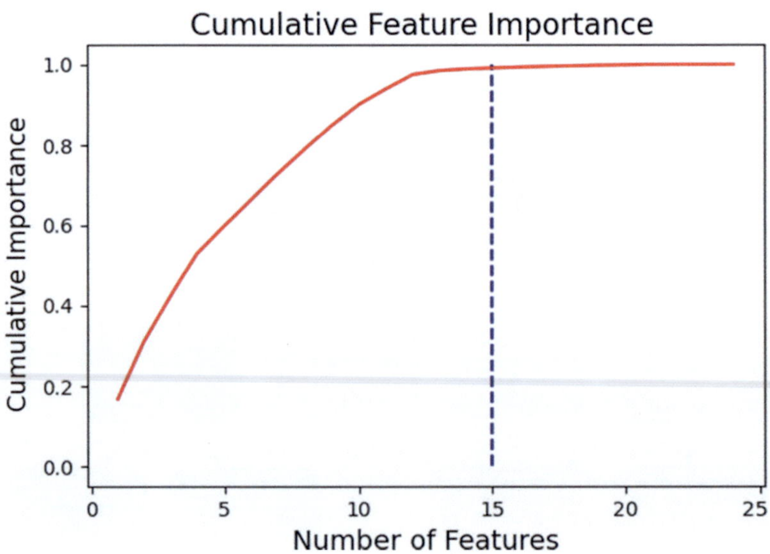

Fig. 5.7 Cumulative feature importance graph

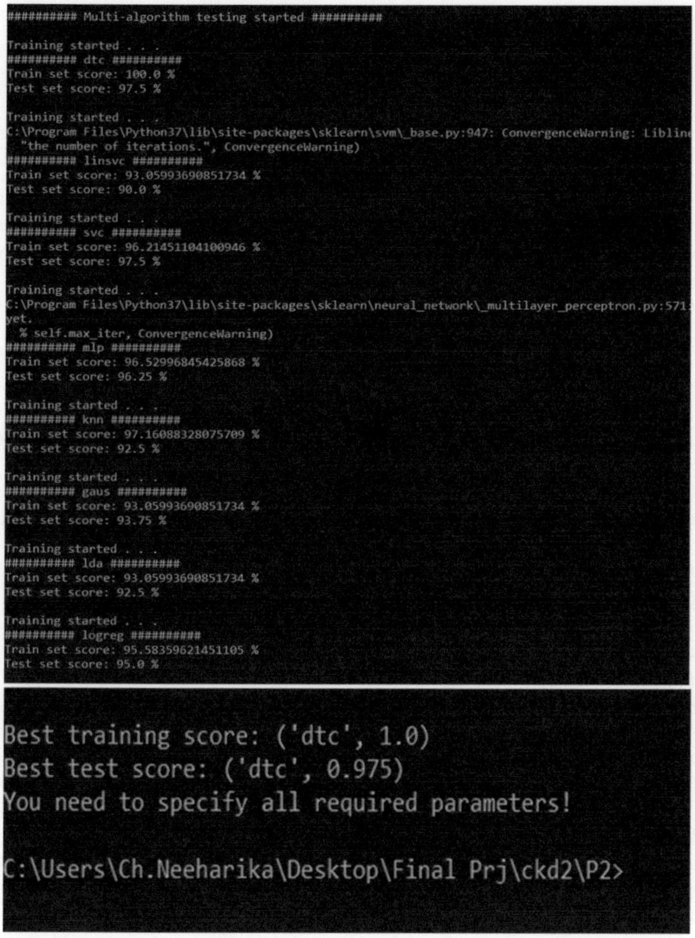

Fig. 5.8 Accuracy comparison of algorithms

Future Work

As the data utilized in this examination is little, later on, we intend to approve our outcomes by utilizing enormous dataset or think about the outcomes utilizing another dataset that contains similar highlights. Additionally, to minimize the predominance of chronic kidney disease, we intend to foresee if an individual with CKD chance factors, for example, diabetes or ancestry will have chronic kidney disease later on or not by utilizing fitting dataset.

References

1. Jha, V., et al.: Chronic kidney disease: global dimension and perspectives. The Lancet **382**(9888), 260–272 (2013). https://doi.org/10.1016/S0140-6736(13)60687-X
2. Scottish Intercollegiate Guidelines Network (SIGN), Diagnosis and management of chronic kidney disease: A national clinical guideline (2008)
3. Kerr, M., et al.: Estimating the financial cost of chronic kidney disease to the NHS in England, Nephrol. Dialysis Transplantation **27**(3), iii73–iii80 (2012). https://doi.org/10.1093/ndt/gfs269
4. Ruiz-Arenas, R., et al.: A summary of worldwide national activities in chronic kidney disease (CKD) testing. Electr J Int Feder Clin Chem Lab Med **28**(4), 302–314 (2017)
5. Dmitrieva, O., et al.: Association of anaemia in primary care patients with chronic kidney disease. BMC Nephrol **14** (2013). https://doi.org/10.1186/1471-2369-14-24
6. Ghosh, D., Vogt, A.: Outliers: an evaluation of methodologies. In: Proceeding on Section on Survey Research Methods—Joint Statistical Meetings, pp. 3455–3460. American Statistical Association, 2012
7. Padmanaban, K.A., Parthiban, G.: Applying machine learning techniques for predicting the risk of Chronic Kidney Disease. Indian J Sci Technol **9**(29) (2016)
8. Salekin, A., Stankovic, J.: Detection of chronic kidney disease and selecting important predictive attributes. In: IEEE International Conference on Healthcare Informatics (ICHI), 2016
9. Tran, C.T., et al.: Multiple Imputation and Ensemble Learning for Classification with Incomplete Data. Springer International Publishing, pp. 401–415 (2017)
10. Schafer, J.L.: Multiple imputation: a primer. Stat. Methods Med. Res. **8**(1), 3–15 (1999). https://doi.org/10.1177/096228029900800102
11. Yildirim, P.:Chronic kidney disease prediction on imbalanced data by multilayer perceptron: Chronic kidney disease prediction. In: IEEE 41st Annual Computer Software and Applications Conference (COMPSAC), 2017
12. Madhuravani, B., Sri Sowmya, G., Sowjanya, P.: Automatic prediction of silent heart attacks using neural networks based hybrid classification system. Int J Adv Trends Comput Sci Eng **9**(5), 7165–7169, 40
13. Abhijith, S., Devika, P., Venkat Siva, A., Jagadeesh, B.N.: COVID-19 detection a system with security alerts, 12 Dec 2020. Available at SSRN: https://ssrn.com/abstract=3747594 or https://doi.org/10.2139/ssrn.3747594
14. Madhuravani, B., Priya Degala, D., Anjaneyulu, M., Dhanalaxmi, B.: Prediction exploration for coronary heart disease aid of machine learning. Turkish J. Comput. Math. Edu. **12**(9), 312–331 (2021)

Chapter 6
Monitoring Suspicious Discussion on Online Forum

P. Srinivasa Reddy, Vijay Keerthika, K. Sai Prasad, and B. Anand Kumar

Introduction

Presently nowadays, individuals are utilizing long range interpersonal communication destinations for correspondence medium. They are getting a huge space for discussion about their stuff with the individual client in many ways like using videos, images, text format form and for the people who are getting those stuff facing the problem like getting confused of whom to choose among a group of cluster and whom they can trust that is who are trusted dealers among the cluster and lot more. So in this way, clients are facing a lot of problems regarding this. For suspect such illicit posts that are content for criminal examination, the law requirement offices are discovering solutions. As most of them are content-based, the proposed framework will concentrate just on content posts.

Checking suspicious talks on online discussion by information mining. The system utilized is information mining in which not much information is removed from an immense measure of information. The framework utilizes content mining to separate suspicious words from the whole visit. The framework gives the discussion to talk as well as lessens the utilization of illicit words during the visit and gives the database to criminal examination if any wrongdoing happened by the individual utilizing that gathering. On the off chance that the framework identifies the suspicious word in the talk, the word is supplanted and if this happens thrice the client record is obstructed for 24 h and if record hinders multiple times, the client will be out of the discussion for all time.

P. S. Reddy (✉) · V. Keerthika · K. S. Prasad · B. A. Kumar
MLR Institute of Technology, Hyderabad, India
e-mail: psrinu6634@gmail.com

V. Keerthika
e-mail: vijaykeerthika@mlrinstitutions.ac.in

V. Garcia Diaz and G. J. Rincón Aponte (eds.), *Confidential Computing*, Advanced Technologies and Societal Change,
https://doi.org/10.1007/978-981-19-3045-4_6

Literature Survey

This section focuses on a variety of connection works that have already been drained in this house. Masses sort of analyzers has contributed their efforts throughout this necessary analysis field. This is {often|this can be} often as follows-

- Allman [1] projected that in the world, the total number of emails that were sent by all of us to all around the world are around 306.4 billions in 2019. And in them out of every fifty emails, there are around five–six mails in the worldwide were spam. This junk mail or spam is in contrast to the law at a lower place of current laws, but can spam disagree with legitimate advertising? If we tend to urge pleasure from observation network TV, employing a social networking Website or checking stock quotes online, we tend to all or any grasp that we might be subjected to advertisements, many of which could be unsuitable or perhaps annoying to the U.S. [2, 3]. Most of the precious shopper services, like social networking, news, and email, are supported entirely by advertising revenue. Whereas people would possibly in distinction to advertising, most customers accept that advertising is additionally a value they purchase accessing valuable content and services. This uninvited industrial email imposes a negative trust on customers with none market mediate profit, and whereas not the likelihood to opt-out.

Bergholz et al. [4] projected that over the previous years, information the information} from collections of photos to genetic knowledge, and to network track statistics are been held on by stylish technologies forming Brobdingnagian datasets. The ever-growing sizes of the knowledge sets have created it crucial to vogue new algorithms capable of handling this knowledge through extreme efficiency [5]. One in all the essential procedure primitives for managing these massive datasets is that the nearest neighbor (NN) problem [6, 7]. The goal is to preprocess a bunch of objects, provides an issue object, and one can understand efficiently the information object most nearly similar to the question [8, 9]. This approach choices a broad set of applications in data analysis and method. As an example, it forms the thought of a largely used classification technique in machine learning: to provide a label for a replacement object, so the foremost similar labeled object and duplicates its label sort of the applications perform data retrieval, search image databases, and understand duplicates and put together sites and many others. Geometric notions are accustomed to represent the objects and their similarity measures [10].

Ho [11] has estimated that phishing emails which are fraudulent practice have usually contain message from a prestigious company or a site which request a user to click on a link to an Internet Website where the user is asked to fill in a form consisting of user details or a completely different counseling [12, 13]. Most of these phishing emails main aim is to retreat money from users or getting access to their personal details. Phishing has increased vastly over the past years and which might be a very serious threat to world security and the economy. There is an enlargement of capable counter measures to phishing [14]. These consists of mathematical theory models for the low-dimensional descriptions of email topics, continuous analysis of email text and links, the detection of embedded logos all these put together can be indicators for hidden seasoning [15]. It is a method which does internal addition or distortion of content which is not perceivable by the reader. For empirical research, we collect a massive realistic corpus of emails which are pre-labeled as spam, phishing, ham words which are legitimate [16]. Finally, filtering of words would possibly be updated and tailored to new kind of phishing.

Proposed System

Data mining is a technique which is much familiar in monitoring social media platforms, further discussion forums for the feedbacks or comments containing any suspicious activity or message. Many of the discussion forums are accustomed to spread any message to an outsized population very instantly. Many people share their views, opinions, and ideas on various topics like politics, religion, and there are few which intentionally hurt sentiments in religious or racial way through malicious posts. So it becomes important to watch posts on such forums. During this paper, we selected a set of set from different online forums. The selected data are converted into a CSV file. On the other hand in neighborhood of this application, if a user logins in to his account and might start a discussion on any topic, if he/she makes use of such illegal or suspicious words, the admin is notified about the particular site also the user is warmed about the same

Figure 6.1 shows us the architecture of the system [6]. It contains steps such as identification and removal of stop words like the, an, a, etc., sentiment analysis for the identification of the suspicious words (Fig. 6.2).

System Design

A use case diagram is a part of unified modeling language (UML), and it created by selecting behavior of the system or the application and by analysis from a use case. It gives us a graphical summary of the mechanism or work done by the system in terms of actors, goals of actors which are also known as use cases and any dependencies between these use cases. The prominent propose of this use case diagram is to give us a clear idea about how a system functions.

Evaluation

Screenshots

In Fig. 6.3, we have the dataset format, which contains the email details like subject and the message in the mail. In Fig. 6.4, we have our application, in which

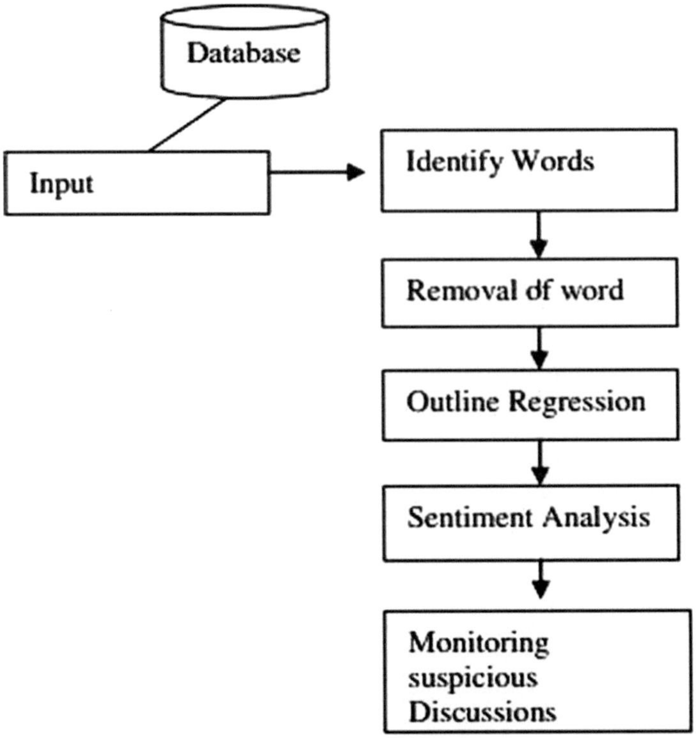

Fig. 6.1 Architecture of the system

we have different buttons assigned to do different tasks. Click on 'upload dataset' button to load forum data.

Figure 6.5 shows us the uploaded data which contain suspicious words, numeric values, and stop words.

We can see all numeric values remove from first and remaining rows. Now, click on 'data stemming' to remove stops words such as off, the, where, and why. Now, click on 'features extraction and generate SVMPSO model' button to generate training mode (Fig. 6.6).

Figure 6.7 shows the prompt for successful generation of model.

Now, click on 'detect suspicious word' button to detect all forums which contains suspicious words (Fig. 6.8).

Figure 6.9 shows us the emails containing the suspicious words in the uploaded data.

Figure 6.10 shows the graphical representation of suspicious chats detected in the uploaded dataset. It is bar graph which has comparison of suspicious chats and total numbers of chats. In graph, x-axis represents total and detected chats, and y-axis represents count. In above graph, total of 30 chats are there and out of that 5 contains

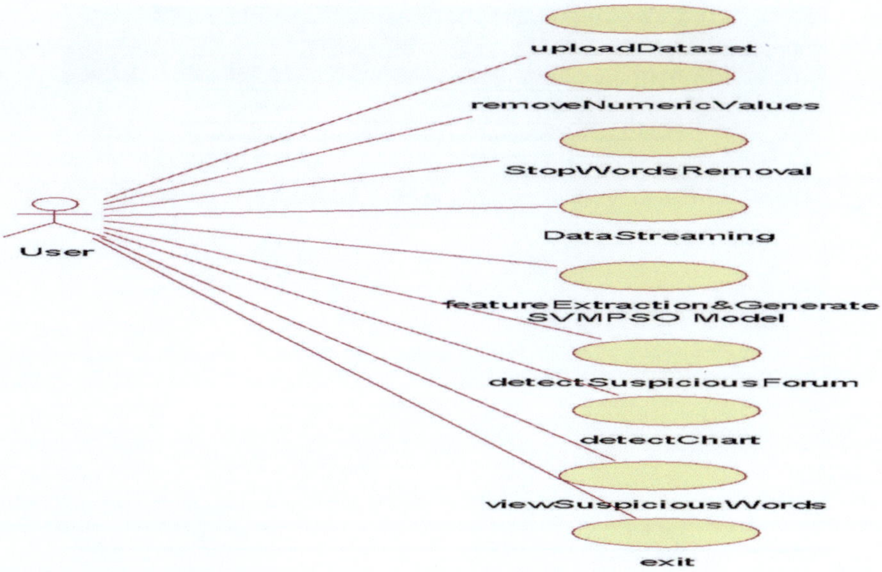

Fig. 6.2 Use case diagram

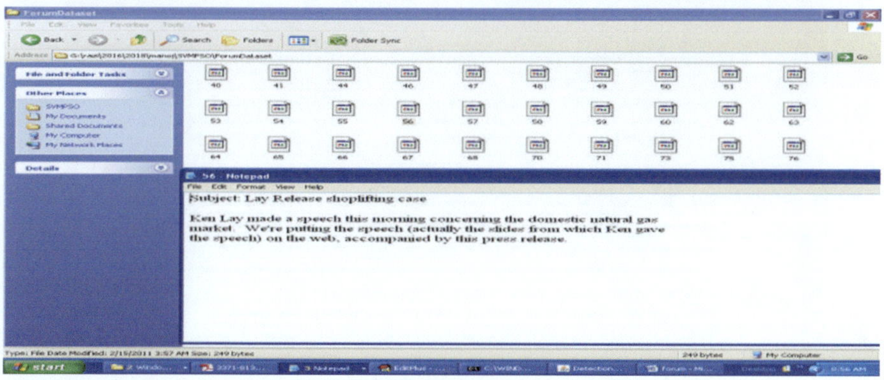

Fig. 6.3 Details of subject and message dataset format

suspicious words. Below screen (Fig. 6.11) showing suspicious words used in this project. This shows us the list of suspicious words detected in the chats.

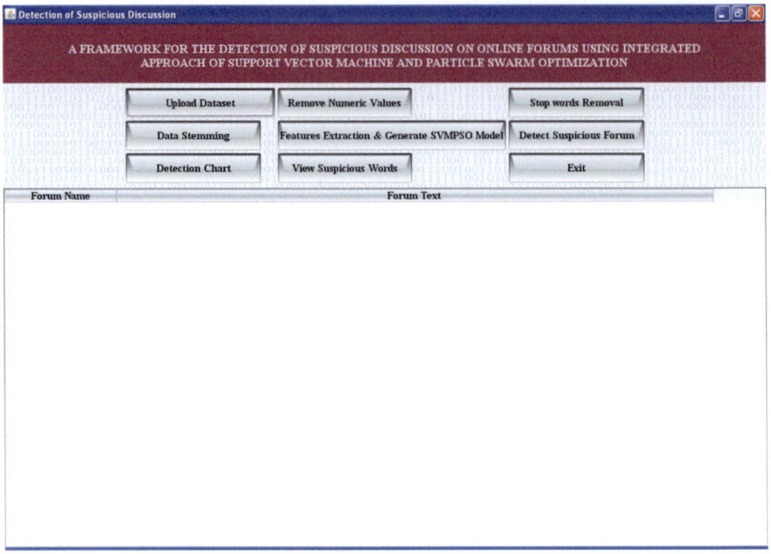

Fig. 6.4 Buttons of different tasks

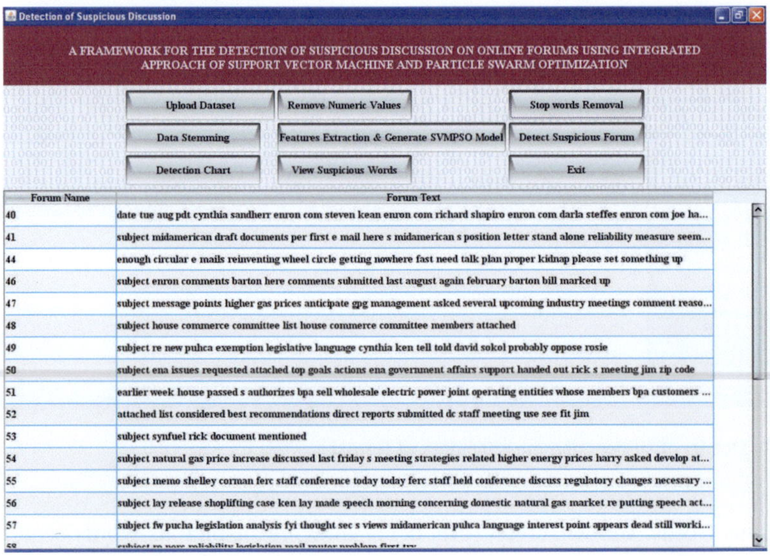

Fig. 6.5 Uploaded data

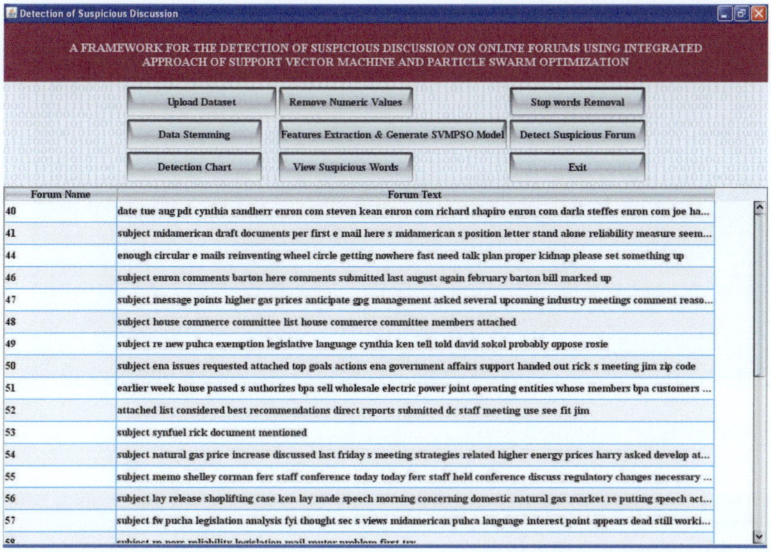

Fig. 6.6 Generated training model

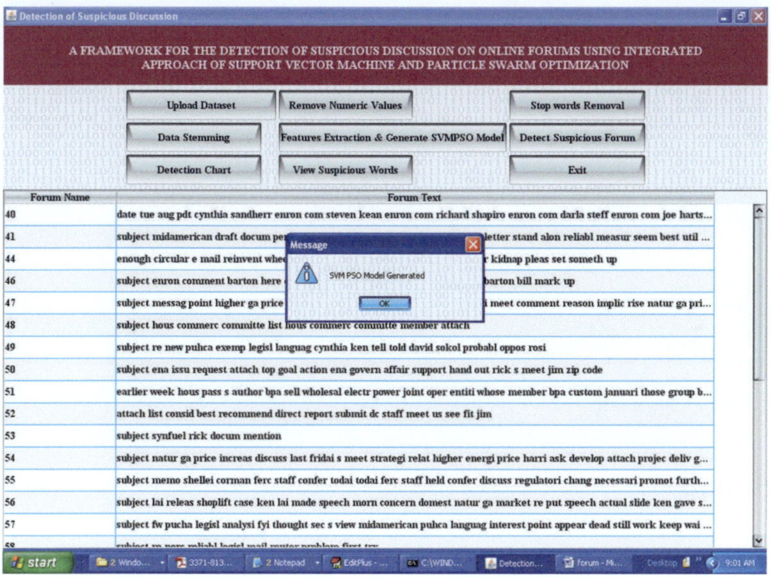

Fig. 6.7 Model of successful generation

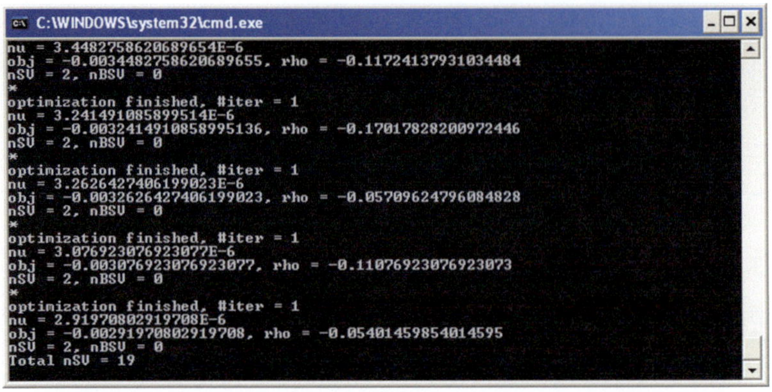

Fig. 6.8 Detecting the suspicious word

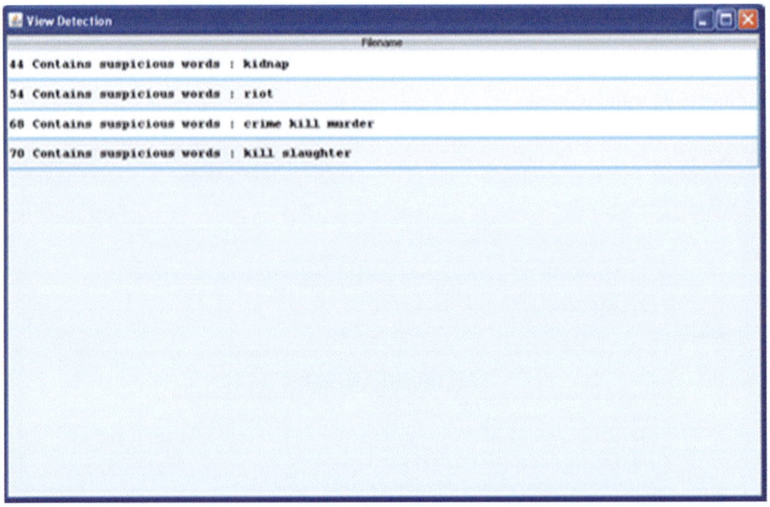

Fig. 6.9 Suspicious words

Conclusion

The main objective of this project is to observe the suspicious activity that happens mostly in online forums.

This application looks from the time the user has logged-in, it checks if any suspicious activity that happens in the online forum area on any topic. If any suspicious words or illegal words are found, then the system replaces it with '*', and a notification is sent to the Website administrator. Therefore, this method detects the suspicious words successfully and prevents such suspicious activities which might cause harm to human life. This application can be in many departments where there is a need.

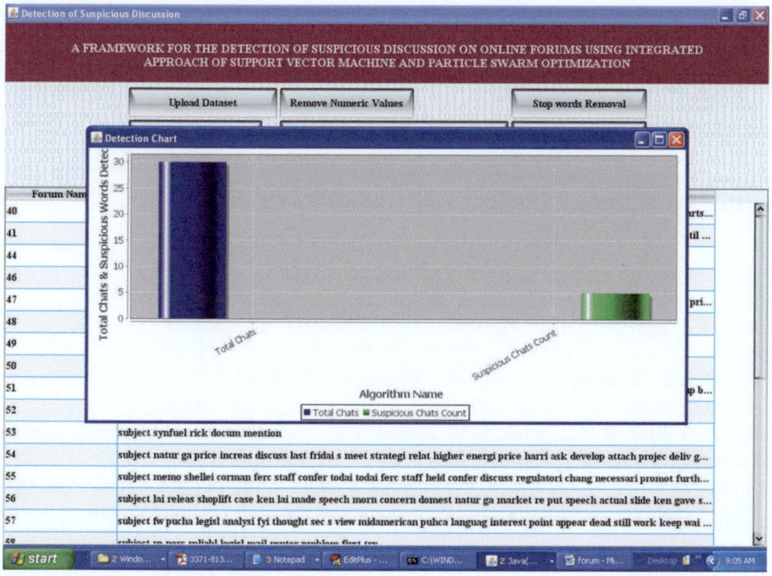

Fig. 6.10 Graphical representation

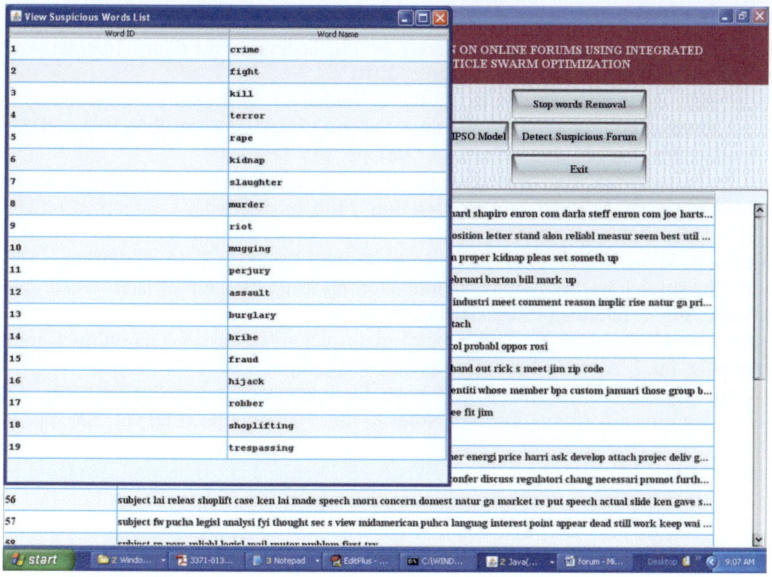

Fig. 6.11 Detected suspicious words

Not only we can use this in social networking sites but it can also be applicable in government departments like forest department, disaster management system to prevent extrajudicial activities.

References

1. Allman, E.: The economics of spam. Queue **1**(9), 80 (2003)
2. Andoni, A.: Nearest neighbor search: the old, the new, and the impossible (2009)
3. Barford, P., Yegneswaran, V.: An inside look at botnets. In: Malware Detection, pp. 171–191 (2007)
4. Bergholz, A., De Beer, J., Glahn, S., Moens, M.-F., Paaß, G., Strobel, S.: New filtering approaches for phishing email. J. Comput. Secur. **18**(1), 7–35 (2010)
5. Bratko, A., Filipic, B., Cormack, G.V., Lynam, T.R., Zupan, B.: Spam filtering using statistical data compression models. J. Mach. Learn. Res. **7**, 2673–2698 (2006)
6. Caballero, J., Grier, C., Kreibich, C., Paxson, V.: Measuring pay-per-install: The commoditization of malware distribution. In: Proceedings of the 20th USENIX Security Symposium, 2011
7. Suruthi Murugesan, M., Pavitha Devi, R., Deepthi, S., Shri Lavanya, V., Annie Princy, Dr.: Automated monitoring suspicious discussions on online forum by data mining. Imp. J. Interdiscip. Res. **2**. ISSN: 2454-1362
8. Porter, M.F.: An algorithm for suffix stripping. Program **14**(3), 130–137 (1980)
9. Connor, B., Balasubramanyan, R., Routledge, B.R., Smith, N.A.: From tweets to polls: linking text sentiment to public opinion time series. In: Proceedings of the Fourth International AAAI Conference on Weblogs and Social Media, 2010
10. Frantzi, K.T., Ananiadou, S., Tsujii, J.: The C-Value/NC-value method of automatic recognition for multi-word terms. In: Proceedings Second European Conf. Research and Advanced Technology for Digital Libraries (ECDL '98), pp. 585–604 1998
11. Ho, T.K.: Fast identification of stop words for font learning and keyword spotting. In: Proceedings of Document Analysis and Recognition, Fifth International Conference on (ICDAR), pp. 333–336. IEEE, Sept. 1999
12. Murugesan, M.S., Pavitha Devi, R., Deepthi, S., Shri Lavanya, V., Annie Princy.: Automated monitoring suspicious discussions on online forums using data mining statistical corpus based approach. Imp. J. Interdisip. Res. (IJIR) **2**(5) (2016)
13. Upganlawar, H., Sambhe, N.: Surveillance of suspicious discussions on online forums using text mining. Int. J. Adv. Electron. Comp. Sci. **4**(40), April (2017)
14. Merugu, S., Reddy, M.C.S., Goyal, E., Piplani, L.: Text message classification using supervised machine learning algorithms. Lecture Notes in Electrical Engineering, vol. 500, pp. 141–150 (2019)
15. Hosseinkhani, J., Koochakzaei, M., Keikhaee, S., Naniz, J.H.: Detecting suspicion information on web crime using crime data mining techniques. Int. J. Adv. Comp. Sci. Inf. Technol. (IJACSIT) **3**(1), 32–41 (2014)
16. Pushpa Rani, K., Jhansi, M., Chandrasekhar Reddy, T.: Best keyword cover search using keyword-NNE algorithm. Int. J. Mech. Eng. Technol. **8**(7 July), 37–43 (2017)

Chapter 7
Ergonomically Designed System for License Plate Recognition Using Image Processing Technique

Divya Priya Degala, M. Anjaneyulu, and P. Devika

Introduction

By using our technology, we are using image processing to stumble the vehicle plate. LPR is employed at several places like malls, parking heaps, etc. This facilitates us to make sure the security. But it is terribly difficult once the background and number plate are analogous [1].

The proposed method makes use of morphological ROI operations to extract the vehicle plate. We can distinguish between background and number plate through the use of morphological operations [2, 3]

Morphological operations are classified into erosion, dilation, opening, and closing.

(1) **Erosion**: The worth of the output constituent is that the minimum worth of all pixels. It may be delineate as if any of the close constituent values is zero, then the pixel cost is about to zero.
(2) **Dilation**: The value of the output pixel is that the most worth of all pixels. It may be outlined as though any of the neighboring pixel values are one, then the pixel cost is about to 1.
(3) **Opening**: It can be outlined as removing objects from foreground and places in background. In beginning, erosion procedure is observed by means of dilation [4, 5].
(4) **Closing**: It may be defined as doing away with small holes in foreground and converts them from background to foreground. In closing, dilation is observed with the aid of erosion.

D. P. Degala (✉) · M. Anjaneyulu · P. Devika
Department of Computer Science and Engineering, MLR Institute of Technology, Hyderabad, India
e-mail: divya.degala@gmail.com

V. Garcia Diaz and G. J. Rincón Aponte (eds.), *Confidential Computing*, Advanced Technologies and Societal Change,
https://doi.org/10.1007/978-981-19-3045-4_7

At the start, the input image is preprocessed by means of using neighborhood maxima-minima binarization technique, observed by means of threshold segmentation then character recognition.

Literature Survey

The motive of this paper is to confirm security. This manner is split into three foremost steps together with processing, segmenting, and recognition of vehicle plate. For this analysis, input image of a vehicle is taken. Here, the output suggests the GUI of registration code [6, 7].

Preprocessing

Binarization approach is manner of adjusting colored image to black and white image. In binarization method, worth of every pixel is calculated. Pixel value stages from zero to 255.

$$\text{No. of. pixels} = \text{pixel columns} * \text{pixel rows}$$

Here, we use local maxima and minima to calculate threshold value. In this technique, threshold measure is calculated via considering most and stripped-down pixel values within the image. By the usage of the non-heritable threshold value, the given input image is remodeled to black and white [8, 9].

Character Segmentation

Initially, it extracts the ROI from whole image and performs threshold segmentation that separates background and foreground. Here, number is taken into account as foreground.

In [10, 11] developed an license plate recognition depending on learning protocol. In this protocol, camera captures the image and provides output as candidate region. The two TDNS which are using as the horizontal and vertical filters provide the license plate.

In [12, 13] suggested a device which will give high-quality images of license plate. In this technique, they used a dual camera developed by author. One camera is stationary and another is pan tilt zoom camera. And an CNN classifier is used to recognize license plate.

Sahas Tabriz et al. present a techniques for recognizing license plate exactly using K-nearest neighbor calculation and SVM techniques [14].

According to Sandipan chowdhary et al. to find out the characters from the dataset and extract the license plate.

According to Muthusamy et al. [15] developed an self-synthesized function of CNN which distinguishes the vehicle plate by its state [16].

Proposed a technique to recognize the license plate by using convolutional neural networks. It uses 53 convolutional model layers, and in second stage, it will do image segmentation for recognition of characters.

According to proposed an ANPR method. In this method, they use two candidate detection algorithms, one is for matching edges and other is for matching the template.

According to Devika et al. [8] developed an ALPR system. In this method, they used two stage identification of characters using simple data augmentation technique.

Proposed Work

There are two principle plate identification strategies utilizing morphological activities: top-hat technique and base-hat strategy. These activities can adequately discover progressively obvious edge areas. Since these techniques are utilize do distinguish edges, they can cause errors if there are solid edges out of sight.

To tackle this issue, we propose the LPR strategy utilizing MROI map as appeared in Fig. 7.1. MROI map comprises of the standard deviation of morphological open and close image. It uses to discover to the brilliance contrast between the background in the plate and the characters. It is imperative to distinguish the plate locales that have exclusive expectation deviations and furthermore are plainly recognized from the background (Fig. 7.2).

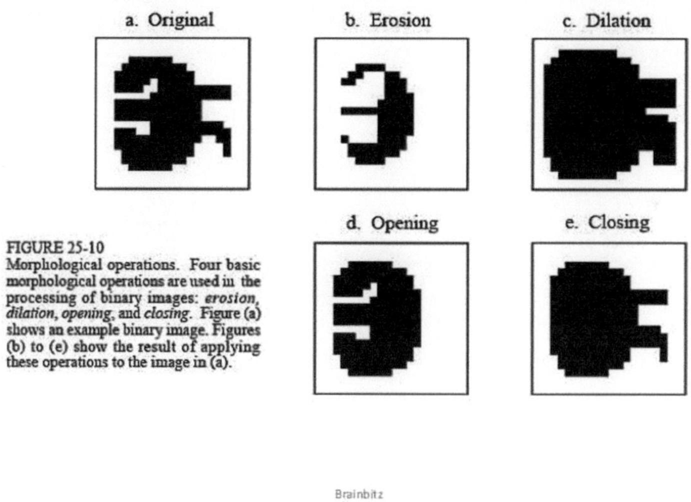

Fig. 7.1 Morphological operations

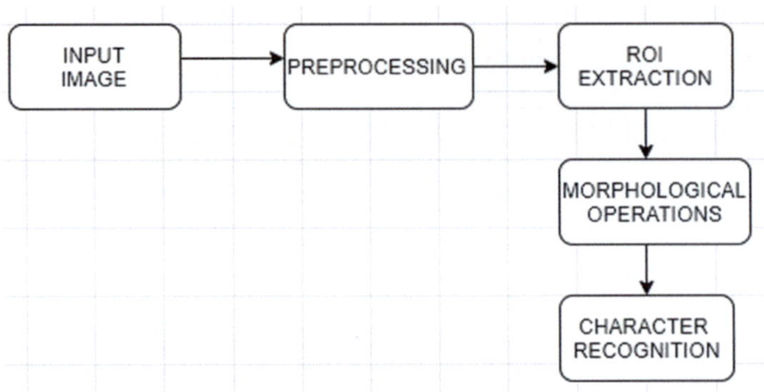

Fig. 7.2 Block diagram

Input Image

A. **RGB Image**

An RGB image is usually spoken as a real color image. It is a 3 dimensional channel with eight bits of red, 8 bits of green, and eight bits of blue (Figs. 7.3 and 7.4).

B. **Grayscale Image**

A gray level image is solely one during which the sole colors area unit reminder gray. The rationale for differentiating such pictures from the other variety of color image is that less data need to be supplied for every picture element (Fig. 7.5).

$$Y = 0.299r' + 0.587g' + 0.114b'$$

Fig. 7.3 RGB image

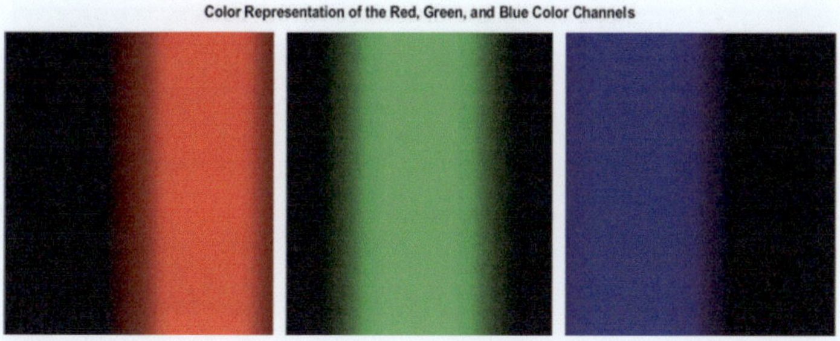

Fig. 7.4 RGB channels

Fig. 7.5 Grayscale image

Preprocessing

A. **Image Acquisition**

This includes filtering a report and putting away it as a image.

B. **Binarization**

Image binarization is that the manner of taking a grayscale image and changing it to black and white, by considering the threshold worth as 127 that is half of 255.

Each pixel value levels from 0 to 255.

Black=0, white=255

Initially, the input image is processed by play acting binarization. Initially, it calculates each pixel value.

Each pixel worth ranges from zero to 255.127 is taken into thought as threshold value. If (pixel value>127)

Converts to white

Else

Converts to black

But if we have a tendency to restoration the threshold value, the results might not be true.

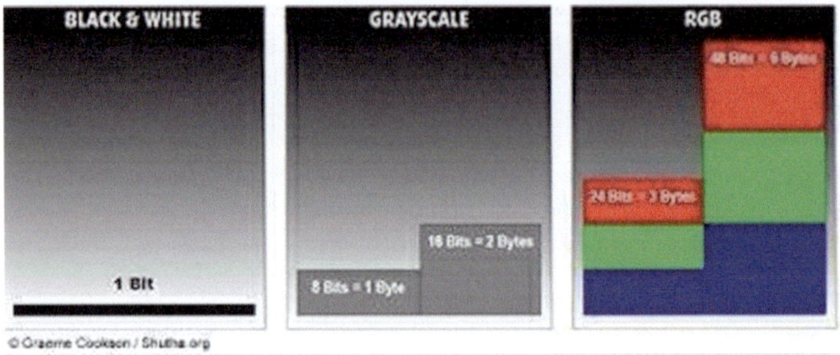

Fig. 7.6 Binarized image

Here, we use local maxima and minima to calculate value of the threshold. In this approach, threshold value is calculated with the help of brooding about most and minimum pixel values at intervals of image (Fig. 7.6).

C. **Connected Component**

It is a crucial procedure in parallel picture preparing that checks an as of now binarized picture and mark sits pixels into segments dependent on PEL availability (four-associated, typically, eight-associated) [3]. When all gatherings of constituents are resolved, every PEL is marked with an incentive as per the segment to which was allocated.

Removing and marking of different disjoint and associated segments in a picture are fundamental to many robotized picture investigation applications, the same number of supportive estimations and highlights in double articles might be separated [3].

D. **Noise Removal**

It is going to be outlined as smoothening of image by obtaining eliminate dots that has additional intensity. Noise removal is completed when binarization is done. It is going to be administrated for each colored and black-white pix.

Character Segmentation

A. **ROI Extraction**

A ROI may be a part of picture that you simply wish to channel or do some alternative activity on [2]. You outline ROI through developing a binary mask that may be a binary image that is the equal length because the image you wish to technique with PEL that outline the ROI set to one and every one alternative PELS to zero [8].

B. **Morphological Operations**

Dilation and erosion are 2 essential morphological operations. Erosion gets eliminate PELS on item boundaries, whereas dilation provides PELS to the boundaries of gadgets in an photograph [7].

Character Recognition

Segmentation procedures can be extensively ordered into 3 classes:

A. **Explicit Segmentation**
In this, information word picture of a grouping of characters is parceled into sub-pictures of individualist characters, which are at that point ordered [5]. This procedure is named as a dissection.

B. **Implicit Segmentation**
It is additionally called recognition-based segmentation. It is methodology is to part words into fragments that ought to be characters, and afterward pass every section to a classifier. On, off chance that the characterization results are not palatable, call division again with the input data about dismissing the past result.

C. **Holistic Approaches**
It is otherwise called segmentation free methodology.
By utilizing this methodology, one can extricate the whole info as apart from a string [6]. This methodology legitimately worry with words, not letters. Utilization of this methodology is restricted to a predefined lexicon.

The OCR uses correlation technique for the character recognition.

Results

See Figs. 7.7, 7.8, 7.9, 7.10, and 7.11.

By seeing above output, we clearly understood that if we take an image of nameplate then with the help of our algorithm, it will convert into string.

Fig. 7.7 Colored image

Fig. 7.8 Colored image to grayscale

Fig. 7.9 ROI extraction of binarized image

Fig. 7.10 Character recognition

Fig. 7.11 Output

Conclusion

Based on our algorithm, the image processing to the vehicle license plate is implemented. By the above examples, we can draw a conclusion that our algorithm can easily recognize the license plate, and recognition accuracy is high. It is used for several functions like tollway regime use this device for permitting the car to travel into the road by way of detection their quantity plate mechanically and supply them with payslip once that open the road for that specific car. Parking government additionally uses this appliance for permitting the car to park of their space. It facilitates us to ensure the security.

References

1. Optasia Systems Pte Ltd.: The World Leader in License Plate Recognition Technology. www.singaporegateway.com/optasia. Accessed 22 Nov 2008
2. Caner, H., Gecim, H.S., Alkar, A.Z.: Efficient embedded neural network- based license plate recognition system. IEEE Trans. Veh. Technol. **57**(5), 2675–2683 (2008)
3. Bellas, N., Chai, S.M., Dwyer, M., Linzmeier, D.: FPGA implementation of a license plate recognition SoC using automatically generated streaming accelerators. In: Proceedings 20th IPDPS, Nice, France, Apr. 2006, pp. 8–15
4. Choudhary, A., Rishi, R., Ahlawat, S.: A new character segmentation approach for off-line cursive handwritten words. In: Information Technology and Quantitative Management, pp. 88–95 (2013)
5. Devika, P.: A smart information system for counting people. J. Adv. Res. Dynam. Control Syst. **5**(3) (2018). ISSN: 1943-023X
6. Devika, P., Prashanthi, V.: RFID based theft detection and vehicle monitoring system using cloud. Int. J. Inno. Technol. Exp. Eng. (IJITEE) **8**(4), 737–739 (2019). ISSN: 2278-3075
7. Anitha, G., Devika, P.: Secure data communication using isecL each protocol in WSNs. Int. Electron. J. Pure Appl. Math. **119**(18), 87–95 (2018). ISSN: 1314-3395 (on-line version). http://www.acadpubl.eu/hub/SpecialIssueJune2018
8. Devika, P., Prasanna, Y., Swetha, P., Akhilesh Babu, G.: Uber data analysis using map reduce. Int. J. Recent Technol. Eng. (IJRTE) **8**(4) (2019). ISSN: 2277–3878
9. Kim, K.K., Kim, K.I., Kim, J.B., Kim, H.J.: Learning-based approach for license platerecognition. In: Neural Networks for Signal Processing X. Proceedings of the 2000 IEEE Signal Processing Society Workshop (Cat.No. 00TH8501), vol. 2, pp. 614–623. IEEE (2000)
10. Mondal, M., Mondal, P., Saha, N., Chattopadhyay, P.: Automatic number plate recognition using CNN based self-synthesized feature learning. In: 2017 IEEE Calcutta Conference (CALCON), pp. 378–381). IEEE (2017)
11. Saif, N., Ahmmed, N., Pasha, S., Shahrin, M.S.K., Hasan, M.M., Islam, S., Jameel, A.S.M.M.: Automatic license plate recognition system for Bangla license plates using convolutional neural network. In: TENCON2019–2019 IEEE Region 10 Conference (TENCON), pp. 925–930. IEEE (2019)
12. Agbemenu, A.S., Yankey, J., Addo, E.O.: An automatic number plate recognition system using opencv and tesser ac to cr engine. Int. J. Comp. Appl. **180**, 1–5 (2018)
13. Laroca, R., Severo, E., Zanlorensi, L.A., Oliveira, L.S., Gonçalves, G.R., Schwartz, W.R., Menotti, D.: A robust real-time automatic license plate recognition based on the YOLO detector. In: 2018 International Joint Conference on Neural Networks (IJCNN), pp. 1–10. IEEE (2018)

14. Manasa, M., Devadasu, G., Sangeetha, S.: Power quality enhancement employing evolutionary algorithm based 17 level asymmetrical multilevel inverter. J. Adv. Res. Dyn. Control Syst. **12**(7), 501–512 (2020)
15. Muthusamy, S.K., Prabha, R., Venkata Hari Prasad, G., Madhusudhanan, B.: Integrating privacy preserving in internet of things to ensure data security using vertical partitioning approach. J. Crit. Rev. **7**(14), 37–41 (2020)
16. Naik, M.V., Anasari, M.D., Gunjan, V.K., Kumar, S.: A comprehensive study of sentiment analysis in big data applications. Lecture Notes in Electrical Engineering, vol. 643, pp. 333–351 (2020)

Chapter 8
Blockchain-Based Privacy Securing G-Cloud Framework for E-Healthcare Service

J. Bennilo Fernandes, Venkat Narayan, Pampana Kusuma Sammilitha, Pavan Sai Koundinya, and R. Ramya Krishna

Introduction

Blockchain

Blockchain technological innovation is transfiguring an extensive variety of companies. Forbes' Bernard Marr highlighted a selection of blockchain channels, including amusement [1–3], for instance music live program technique Spotify; the meals company, like for resource chain supplying; and medical, for instance for storage and use of overall health documents [4, 5].

Previous Hash. The prior hash is literally the feature which links a obstruct to the previous block. It is composed of the hash worth of the previous block [1, 6].

Data. The product includes the sender's offer with the receiver's standard address in addition to the transfer length. Right now, there are numerous transactions involving all receivers and senders; thus, each block includes each volume of transactions, and each transaction is able to end up with a sender's standard address and a receiver's regular address, together with a transactional nonce [7, 8].

Nonce. Bitcoin has a proof of methodology, and to be able to do the algorithm, a random printer is really employed varying the output of the hash benefit; this is referred to as the nonce. Proof of work would be the process of transaction validation [9, 10].

Hash. The hash is much like an electronic fingerprint. It seems a thing like a hexadecimal printer. Public Passed out Ledger to recap, a blockchain happens to be really a decentralized public sent out ledger that is actually used to record transactions throughout several PCs and laptops. For example [11, 12], specific A transfers money to computer user B, desktop operator B transfers to C, and also C transfers to B. A

J. B. Fernandes (✉) · V. Narayan · P. K. Sammilitha · P. S. Koundinya · R. R. Krishna
Department of ECE, Koneru Lakshmaiah Education Foundation, Guntur, Andhra Pradesh, India
e-mail: bennij@kluniversity.in

V. Garcia Diaz and G. J. Rincón Aponte (eds.), *Confidential Computing*,
Advanced Technologies and Societal Change,
https://doi.org/10.1007/978-981-19-3045-4_8

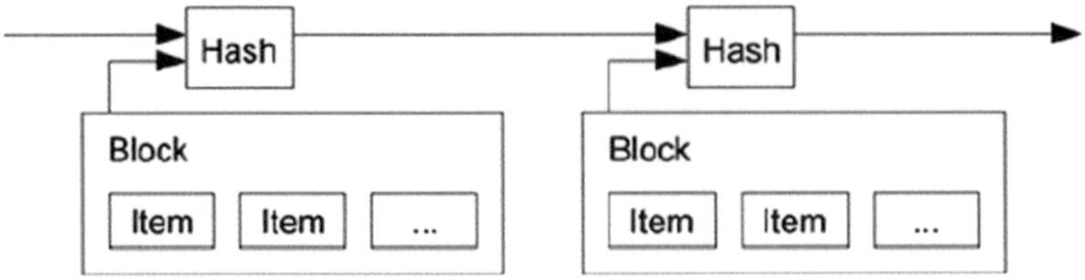

Fig. 8.1 Hash architecture

dispersed ledger might be essentially a data source that is discussed with all of the proprietors which are essentially a portion of the blockchain system that is proven in Fig. 8.1.

Existing System

Today's process for electronic health and fitness article sharing, dependent largely on blockchain network, is actually released. For that particular motive, the blockchain primarily based on EHR revealing process architecture is really recommended. The device workflow is quite simple to work with. Individuals buy via the prospect program or perhaps SDK, requesting an enrollment certification with a club membership system provider (MSP) [13] due to the certification specialist. Then, the license professional problems the certification and individual component with a new ID to enlist the participant. As a consequence of committing the transaction inside the blockchain marketing, the current transfers are in fact sent across the social networking. The vendors, such as clinicians in addition to lab personnel, could query required specifics across the system. If the affected individual grants utilization of amenable and also increase the data of theirs inside the EHR ledger marketing, then the clinician or maybe laboratory participant has the ability to open as well as upgrade whenever required for authorization files of the people. The disadvantages of the current product are pointed beneath.

Proposed System

The proprietors of this specific approach may be management, patients, staff, doctors, etc.. The main job of these individuals is reaching the system and also do very simple tasks like create, look at, update, and delete the info. The alternative coating on the process is in fact the blockchain range; the code is found by this particular level or perhaps possible mechanism for interaction of computer operator with all the DApp which in turn is really operating about the blockchain. This specific level has three components around it. They are actually: Try adding information would grow person's overall health files within the DApp. It is the areas of IPFS, name, co-morbid, and blood group, along with ID hash. The individual's main medical files tend to be literally conserved along with the IPFS hash which has the file published which has the laboratory outcomes or even maybe every other health info of long-suffering. Replace data would increase the health-related files of unhurried. This may simply modify the basic information belonging to the impacted individual not the IPFS hash. IPFS hash is certainly non-changed to help make positive protection of documents. Perspective documents will permit the person see the files of an individual stored in DApp. The perspective record performance is really employed by doctors and patients. The person can see the data of his next to the procedure authenticating affected person opinions merely the very own health-related documents of his. The affected individual to generate sure that only the relevant medical files tend to be essentially established toward the customer.

Module Description

Module 1 Login Screen and Dashboard

To use the Identity-Based Integrity Auditing program, you have to first supply a username as well as password. Authentication working determining a user and confirming this end user is able to use the application is explained in Fig. 8.2.

In order to access your Identity-Based Integrity Auditing application, enter the below URL in your Internet browser site address bar: http://127.0.0.1:9091.

Before being permitted to use the Web site application equipment, you will be directed to login. The login display for the Identity-Based Integrity Auditing Software Administrator asks to get a username as well as password. Your password and

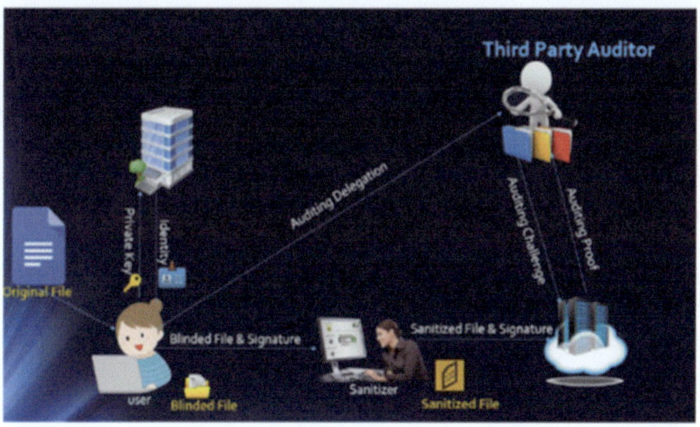

Fig. 8.2 Proposed system work flow

username are going to be supplied by your program user. To login for the application program Admin:

- Enter your Password and Username.
- Click the Remember Me package to remain logged in for up to fourteen days.
- Click the Log In button to get access to the software program administrator.

Module Two Doctor Information

Out system goal is to optimize the skills of hospitals, multi-speciality clinics, medical practitioners, and doctors by automating the tasks of recording patient's info.

Instance Attributes. Objects are created by all classes, and all items have attributes known as characteristics (referred to as qualities in the opening paragraph). Inheritance is actually the procedure by which one category takes on the characteristics and strategies of another. Newly formed classes are actually called kid classes, as well as the classes that kid classes are actually derived from are actually known as parent classes. It is crucial that you should be aware that kid classes override or even lengthen the performance (e.g., behaviors and attributes) of parent classes.

Module 3 Patient and HER

Patient Information. This is the module where in all of the patient group details may be taken care off. An irreversible registration number, which is going to act as a patient identifier, is actually allocated in this case. The registration number offered to the individual will be saving the details of amount of trips to the O.P.D. and selection of admissions in the medical center. Distributed health information networks have been recommended to enhance the capability to obtain and analyze information across institutions leading to enhanced quality, safety, and effectiveness of care.

Electronic Health Record. Healthcare data encryption is actually a type of information protection whereby electronic health records (EHR) are actually disguised so that unauthorized people might not read or even make good sense of them. Private health info (PHI) including health diagnoses and surgeries along with other vulnerable health information must be anchored to guard against malicious motives also as confidentiality breaches which can lead to great fines.

About File Upload. The HTTP strategies specified in the Allow directive are actually the 2 techniques utilized by the HTTP server to successfully pass info to the CGI method (Application Broker). The Make it possible for directive lists the allowable values due to the demand technique; this line does not truly set the method. The method names are actually Get and POST. Get directs the server to process the whole form as a person very long concatenated string of values added toward the URL. Making use of Get enables users to save the resulting powerful web pages.

Module for Blinded File and Sanitized File Generation

- For preserve security of other and financial R&D information with different internal data which really needs to be secure from intruders.
- The organizations or the employees do not want their private or financial things to be made public.
- To make fast transitions so that private things are not made public.

Encryption scrambles text to really make it unreadable by anyone apart from those with the secrets to decode it, and it is starting to be much less of an additional choice and much more of a must have component in any sort of security program for its power to relax and also deter hackers from stealing very sensitive info.

Algorithm for Blinded File

- Python accepts the file suggestions and encrypts it making use of Pycrypto module.
- The filename is actually taken as input parameter together with the password.
- Encryption is actually attained with the assistance of key that is produced with algorithmic standards. The encrypted file is actually stored in the exact same directory with a prefix of (encrypted) included to it.
- Cryptographic hash functionality makes keys which allows in maintaining security of documents that are utilized for symmetric encryption of documents.
- The files are essentially protected with passwords.
- Decryption follows reverse process in which password and encrypted file are actually taken as input parameters.

While lengthier keys provide the owners with much stronger encryptions, the power comes at the price of performance, which means that they are going to take longer to encrypt. The hash feature takes a larger file as feedback, processes it, and returns a smaller result which is practically assured to be a distinctive fingerprint of the file. It is then very easy to compare 2 documents to see whether they are completely different from each other. Actually changing one character is going to result in a certain hash output. Hashing is often used around cooperation with encryption.

Results

To be able to use the Identity-Based Integrity Auditing programs, you have to offer a username as well as password as shown in Fig. 8.3. Authentication means determining a user and confirming that this particular user is actually permitted to use the application. That the is actually the web page where we are going to provide the login information to be able to use the application.

If the individual enters the right login credentials, we are going to land up during these Web site. This is the dash panel belonging to the software program as shown in Fig. 8.4.

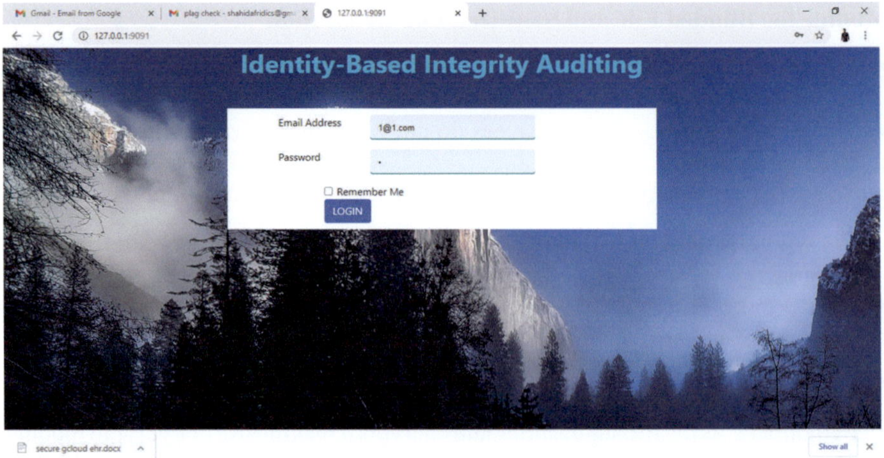

Fig. 8.3 Login page

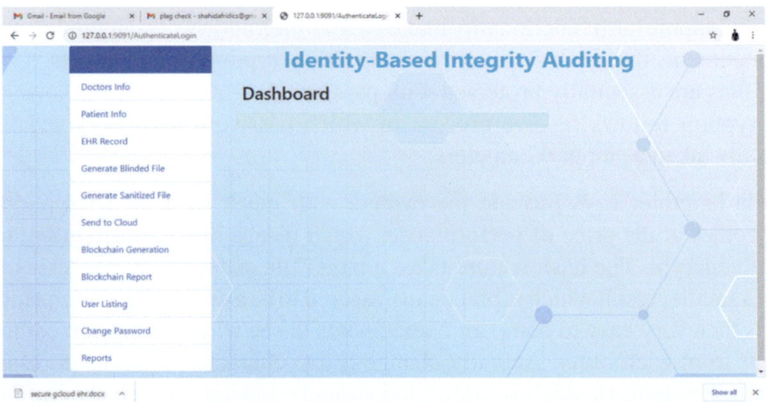

Fig. 8.4 Dashboard

Our system goal is to optimize the skills of hospitals, multi-speciality clinics, medical practitioners, and doctors by automating the tasks of recording patient's info. This web page displays all of the medical doctors within the hospital, we are able to include brand new medical professionals to the summary, and we are able to also alter the info of doctors. We are able to include details like the specialization of his and details are contacted by doctors is shown in Fig. 8.5.

This web page consists of all of the patient details as shown in Fig. 8.6, and we are able to also include new individuals and assign them to certain physician that could help the individual. The goal of collecting as well as storing patient info is actually making it accessible for decision-making at a point of attention or even for action

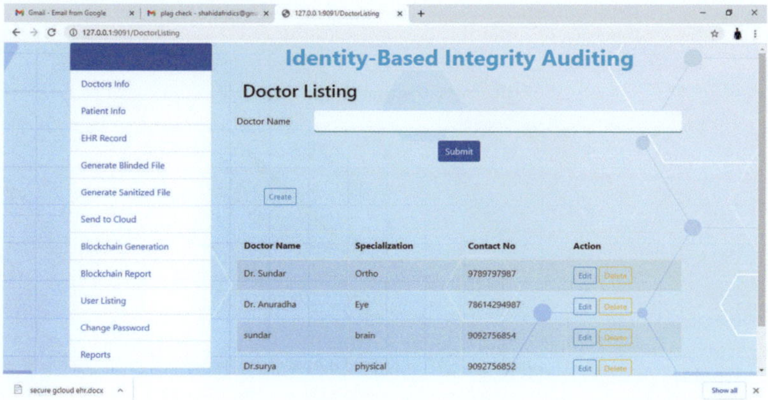

Fig. 8.5 Duty doctor's list and specialization

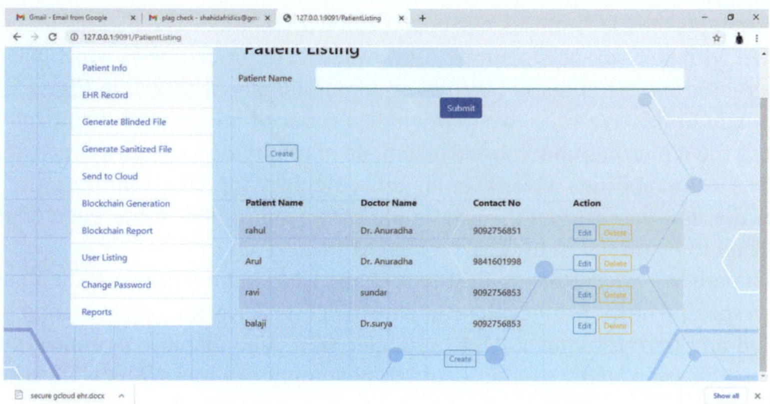

Fig. 8.6 Patient's list

and analysis for policy and management. We are able to often edit or perhaps delete patient info.

The track record has a selection of "notes" sorts got into after some time by total healthcare experts, capturing observations and administration of drugs and therapies, orders for that particular administration of drugs and therapies, reports, X-rays, test results, and much more. The upkeep of correct and complete wellness info is a need of healthcare distributors, and also, it is often enforced like a licensing or perhaps possibly accreditation requirement.

Right here, we are able to information healthcare records associated to the individuals as shown in Fig. 8.7. We are able to actually publish appropriate documents associated to patient here.

In this web page, we are producing blinded file, for the affected person information. The algorithm utilized for generating blinded file is through using python.

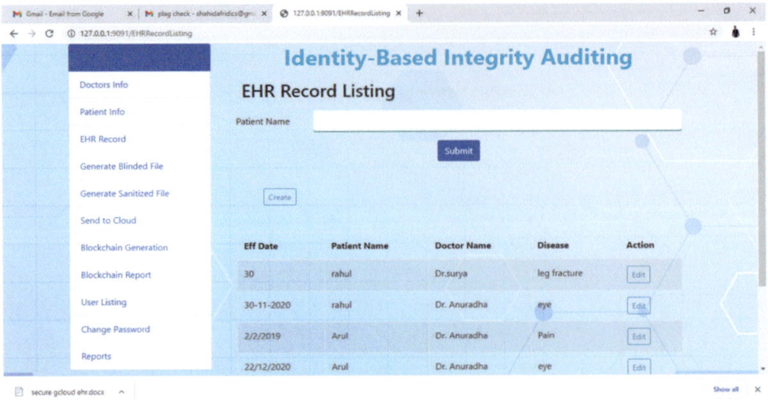

Fig. 8.7 EHR record

Python accepts the file feedback and encrypts it with module. The files are essentially protected with passwords. Decryption follows reverse procedure in which password and encrypted file are actually taken as input parameters as shown in Fig. 8.8.

Throughout this Web site, we sanitize the produced blinded file. Sanitized file is actually a file which has undergone a technique of removing some embedded artifacts as well as vulnerabilities while keeping a file's enhancing. The sanitized file would still get the first data format as well as file extension (unless it has been explicitly established to be turned in to the next kind of file).

Right here, we are producing the block chain. Each patient information is actually kept in a specific block in blockchain as shown in Fig. 8.9, which makes it much more protected and hard to tamper. This particular web page displays complete number blocks which are actually produced and total amount of blocks which are still to be created.

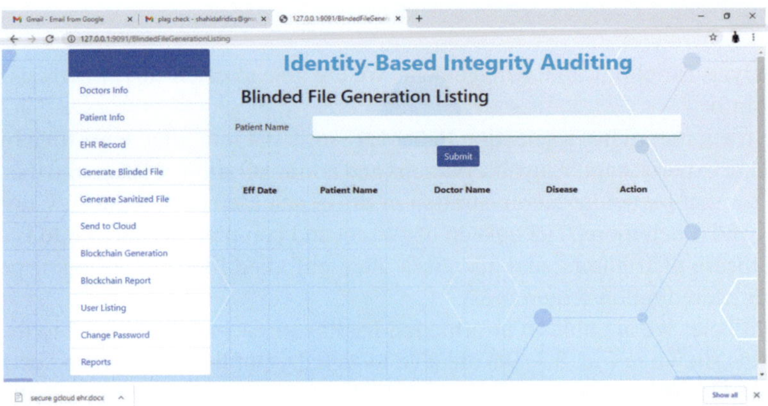

Fig. 8.8 File blinding

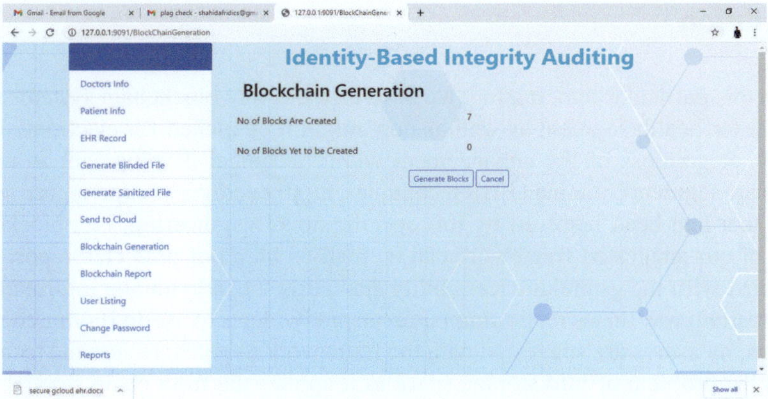

Fig. 8.9 Blockchain generation

The blockchain article is displayed by this particular page. Blockchain eliminates unauthorized entry by using the cryptographic algorithm SHA 256 to be sure that the blocks are in fact stored, protected, and revealed as shown in Fig. 8.10. Each and every person inside the blockchain has his or possibly her keys: a private system as well as a public system. The personal primary factor is in fact well known simply into the sender; it is likewise used to determine if the beginnings on the transaction are really genuine.

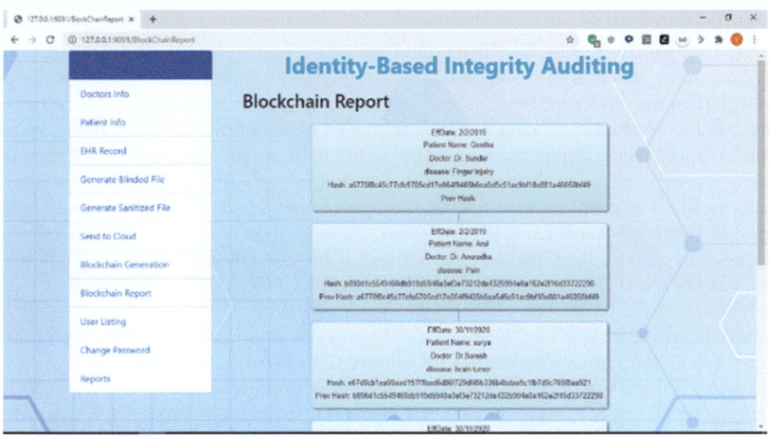

Fig. 8.10 Encrypted blockchain report

Conclusion

Within this particular undertaking, we described the way blockchain systems might be ideal for health segment as well as how might it be picked for electronic health records. Regardless of the enhancement within technical development as well as healthcare segment contained EHR techniques, they nevertheless experienced several issues that had been resolved by this specific novel engineering, i.e., blockchain. Much of our suggested framework can be really a blend of protected report space combined with the granular accessibility standards for individual's information. It makes certain way that is really simpler for any individuals to use and comprehend. In addition, measures are suggested with the framework to generate sure the technique covers the problem of info storage space as it applies the from chain storage space mechanism of IPFS. Also, the job primarily based on accessibility in addition gains the procedure because the overall health files are just provided on the dependable as well as related individuals. Now, this resolves the problem of information asymmetry of EHR procedure.

References

1. Li, X., Jiang, P., Chen, T., Luo, X., Wen, Q.: A survey on the security of blockchain systems. Futur. Gener. Comput. Syst. (2017). https://doi.org/10.1016/j.future.2017.08.020
2. Hoque, N., Bhattacharyya, D.K., Kalita, J.K.: Botnet in DDoS attacks: trends and challenges. IEEE Commun. Surv. Tutorials **17**, 2242–2270 (2015). https://doi.org/10.1109/COMST.2015.2457491
3. Zhang, Y., Deng, R.H., Liu, X., Zheng, D.: Blockchain based efficient and robust fair payment for outsourcing services in cloud computing. Inf. Sci. (Ny) **462**, 262–277 (2018). https://doi.org/10.1016/j.ins.2018.06.018
4. Chiew, K.L., Yong, K.S.C., Tan, C.L.: A survey of phishing attacks: their types, vectors and technical approaches. Expert Syst Appl **106**, 1–20 (2018). https://doi.org/10.1016/j.eswa.2018.03.050
5. Bradbury, D.: The problem with bitcoin. Comput. Fraud Secur. **2013**, 5–8 (2013). https://doi.org/10.1016/S1361-3723(13)70101-5
6. Radanović, I., Likić, R.: Opportunities for use of blockchain technology in medicine. Appl. Health Econ. Health Policy (2018). https://doi.org/10.1007/s40258-018-0412-8
7. Koshechkin, K.A., Klimenko, G.S., Ryabkov, I.V., Kozhin, P.B.: Scope for the application of blockchain in the public healthcare of the Russian Federation. Procedia Comput. Sci. **126**, 1323–1328 (2018). https://doi.org/10.1016/j.procs.2018.08.082
8. Azaria, A., Halamka, J.D., Lippman, A.: A case study for blockchain in healthcare: "MedRec" prototype for electronic health records and medical research data. Proc. IEEE Open Big. Data Conf. **13**, 13 (2016). https://doi.org/10.1006/geno.1994.1313
9. Theodouli, A., Arakliotis, S., Moschou, K., Votis, K., Tzovaras, D.: On the design of a blockchain-based system to facilitate healthcare data sharing. In: Proceedings of 17th IEEE International Conference on Trust, Security and Privacy in Computing and Communications, 12th IEEE International Conference On Big Data Science And Engineering, pp. 1374–1379 (2018). https://doi.org/10.1109/TrustCom/BigDataSE.2018.00190
10. Xia, Q., Sifah, E.B., Asamoah, K.O., Gao, J., Du, X., Guizani, M.: MeDShare: trust-less medical data sharing among cloud service providers via blockchain. IEEE Access **5**, 14757–14767 (2017). https://doi.org/10.1109/ACCESS.2017.2730843

11. Hussein, A.F., ArunKumar, N., Ramirez-Gonzalez, G., Abdulhay, E., Tavares, J.M.R.S., de Albuquerque, V.H.C.: A medical record managing and securing blockchain based system supported by a genetic algorithm and discrete wavelet transform. Cogn. Syst. Res. **52**, 1–11 (2018). https://doi.org/10.1016/j.cogsys.2018.05.004
12. Rao, G.S., Sampath Raju, S.: Multilevel inverter-based power quality improvement in grid-connected DVR system. Lecture Notes in Networks and Systems, pp. 361–366 (2021)
13. Kumar, S.S., Srinivasa Rao, G., Voggu, S.: Fuzzy logic controller with zeta converter for electric vehicles induction motor. Lecture Notes in Networks and Systems, pp. 617–623 (2021)

Chapter 9
Development of Raspberry Pibot Surveillance Security System

Swathi Cherukuri, Ramcharan Chenniboyena, Deepu Yarlagadda, Venkata Ratnam Kolluru, and Shaik Razia

Introduction

In order to protect their houses, people usually use safety cameras. In the storage unit, the cameras capture and store activities. In the case of dangerous or unusual incidents or actions, action can be taken after reviewing the footage. This paper develops a home security system based on the IoT platform [1]. There are several existing devices which are used for domestic defence, such as microcontroller-based wired and wireless security systems and CCTV, but they are very expensive and have limited user range and access. In this research work, a cost-effective home protection system based on Raspberry Pi with PIR sensor was developed and mounted. We implemented a different project compared to other projects. We also did literature survey on different papers where they used different types of methods and techniques for the result [2]. The base paper we used tells about how they used different types of images as inputs where the output will be face detection, so here their main method is face recognition [3, 4]. Second paper tells about how they used smart phone for controlling the access to house where their main aim is also about security [5]. Kotapalle et al. also implemented some work in which they told about how they used machine learning-deep learning algorithms for detections process [6, 7]. Fourth paper tells about how they used automatic control system for controlling the appliances in the home [8]. Kumar et al. implemented a fast and cost-effective face recognition system using CNN algorithm, so main reason I used this model paper is because we also implemented CNN algorithm for a small backend process where image processing is done [9, 10]. Intruder detection with alert using cloud-based CNN and Raspberry Pi is another survey in which they cloud storage and implemented it using deep learning techniques [11, 12]. Seventh paper is about how they just used security

S. Cherukuri · R. Chenniboyena · D. Yarlagadda · V. R. Kolluru (✉) · S. Razia
Department of ECM, Koneru Lakshmaiah Education Foundation, Guntur, Andhra Pradesh, India
e-mail: venkataratnamk@kluniversity.in

V. Garcia Diaz and G. J. Rincón Aponte (eds.), *Confidential Computing*,
Advanced Technologies and Societal Change,
https://doi.org/10.1007/978-981-19-3045-4_9

system to detect buglers and preventing from entering into house using Raspberry Pi [13]. So by basing all this survey, we implemented an idea which is an all-in-one model where we can do everything on a single embedded testbed which includes intruder detection and also can give instructions to our camera to rotate in 180°; main advantage is we can give it through our mobile telegram application, and with all this, we can also see the live surveillance in our own smart mobile phone.

This paper work is carried forward with different measures by explaining. In Sect. "Installation Process for Raspberry Pi", we can know how the process is done and how we installed Raspberry OS; in Sect. "Proposed System for Intruder Detection and Surveillance", we can come across how the development is done using flow graph and block diagram; in Sect. "Functionality of Intruder Detection Model", we can know the knowledge how the working is and also the view about how it looks like and we can use it; in Sect. "Results and Discussions", we can see results and discussions; in Sect. "Conclusion", we can know what is the proper conclusion for this model and how we observed the output; and Sect. "References" finally says us about what are the references we used for the model implementation.

Installation Process for Raspberry Pi

The IDS is intended for cyber antiques planning and care. Systems for detecting intrusions (IDS). This is accomplished by extracting data from a range of sensors and network sources and then evaluating security problems. The proposed approach using the Raspberry Pi board is the main controller. The newest Raspbian update is available on the board. After the OS has been fixed to the frame, connect the hardware and switch on the power supply. To connect the Raspberry Pi, it begins booting the board and username and passwords. It runs on the Debian Linux Operating Arch system. It primarily uses Python software and handles network setup by terminal window commands to upgrade the Python software. The following packages must be installed for the application of the model suggested. For installation commands, the following were listed.

1. sudo apt-get python-matplotlib is installed
2. sudo apt-get python-numpy install
3. sudo python-SciPy apt-get install
4. sudo apt-get python-imaging

To capture and save the picture in the archive, install the OpenCV and camera setup packages on the board. Run the Python code and erase the noise in an image to check for algorithms for enhancement. Implementation is seen in Fig. 9.1 of the proposed process.

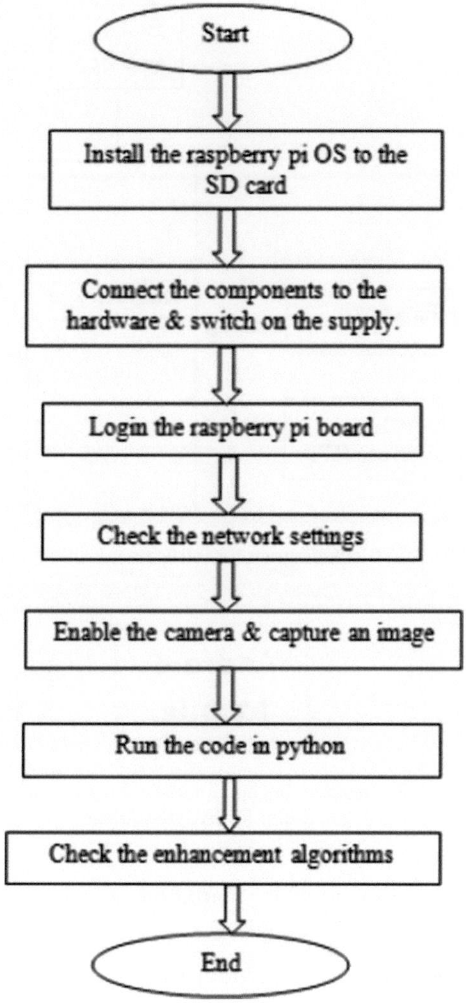

Fig. 9.1 Flow graph of intruder detection process

Proposed System for Intruder Detection and Surveillance

This paper explores the design and deployment of the smart monitoring system with Raspberry Pi and PIR sensors for mobile devices. It requires mobile technology to be used to provide essential protection for our homes and other control applications. The proposed home surveillance device captures and transmits data to the corresponding email through the Internet. Raspberry Pi operates and controls motion detectors and video cameras for remote sensing and monitoring, transmits live video and preserves it for future replication. It is also possible to locate the number of people using the infrared sensor. For example, the Raspberry Pi device informs the

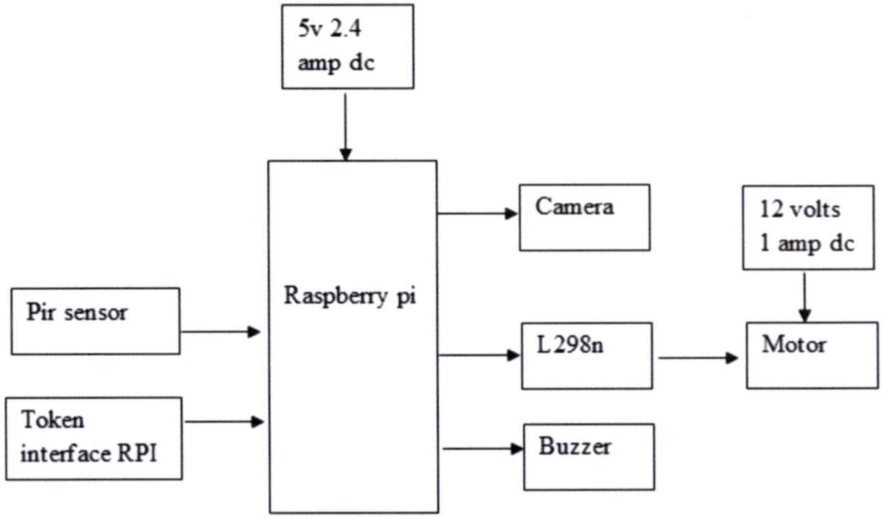

Fig. 9.2 Block diagram for intrusion detection and surveillance system

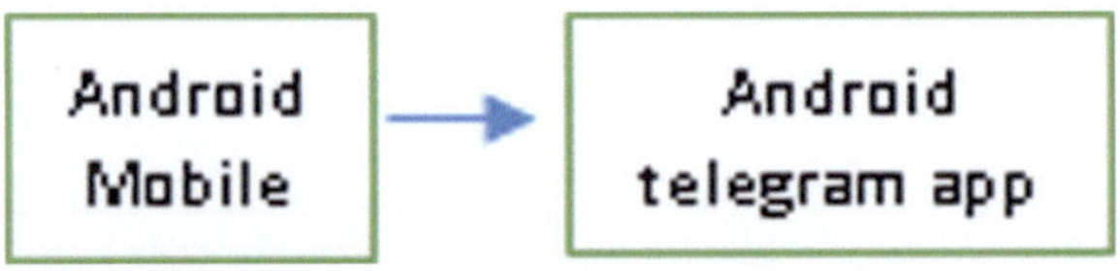

Fig. 9.3 Telegram bot app for controlling the camera for rotating it 180° by giving instructions for camera which is useful for live surveillance

owner of a possible smartphone attack when motion is sensed that the cameras start automatically recording.

Flow graph for intruder detection

See Fig. 9.1.

Block diagram for hardware implementation of intruder detection kit

See Figs. 9.2 and 9.3.

Functionality of Intruder Detection Model

We use the PIR sensor to identify intruders for this project, and the Pi is alerted and collected when the intruder has been identified by the sensor. After the picture has been added and e-mails have been sent to the user, the user will then bind to the login credentials of a telegram programme and pick a bot to give left and right instructions

Fig. 9.4 Intruder detection system hardware setup

to rotate the camera and use those applications for live streaming, where this live streaming can be observed by us in our own smart phone in chrome browser.

Here, the main scope of this project is to ensure the security of home from intruders.

1. Firstly, we will connect our Raspberry and give power supply to it.
2. Next, when we run our intruder code then if the PIR sensor senses anything it will alert the Pi camera and next capture the image and send it to our email using the SMTP protocol which is inbuilt in Raspberry. To this, we connected the motor driver board for 180° rotation of the camera when we wish to see the left and right side view when intruder detected.
3. To perform these operations, we are creating a bot using telegram to perform the above-mentioned operations.
4. So after this, to increase some accuracy to this project, we are including live surveillance; this process is done using chrome browser using Raspberry IP address and TCP port number 8000. Here, port 8000 is used as live streaming port id (Figs. 9.4, 9.5 and 9.6).

Results and Discussions

Raspberry Pi intruder detection kit interfacing the PIR sensor, camera, motor driver, dc motor and buzzer is implemented. When the Raspberry Pi detects the intruder,

Fig. 9.5 Telegram bot app controlling the camera by giving instructions to rotate

Fig. 9.6 Live streaming image observed in chrome browser

it captures image and sends an alert to the user in mail by attaching the captured image to the mail and alert user like intruder detected and please have a look at the live surveillance. When the user wants to access the camera for having left and right 360° view by accessing IoT-based telegram android application.

Using the Raspberry Pi that has a camera attached to it, we confirmed the result, there is also an output verified on embedded domain platform that is observed from Raspberry Pi, and software output is observed in a software tool to get required output.

Conclusion

An IoT-based home security system was developed and built with the Raspberry Pi 3, Pi camera and PIR sensor. The owner will get updates anytime and wherever on a mobile or laptop. It warns if any motion that is unfamiliar or suspicious is detected. This model blocks every unauthorised entity from successfully entering the building, where it can all be set for the smart home world. For easier use and monitoring of the machine by the elderly and people with disabilities, the Raspberry Pi cam telegram controlled cam rotation software is available. This project includes a basic framework for home security that is simple to install and use.

References

1. Somani, S., Solunke, P., Oke, S., Medhi, P., Laturkar, P.P.: IoT based smart security and home automation. In: 2018 Fourth International Conference on Computing Communication Control and Automation (ICCUBEA), Pune, India, 2018, pp. 1–4. https://doi.org/10.1109/ICCUBEA.2018.8697610
2. Kishore Kodali, R., Rajanarayanan, S.C., Boppana, L., Sharma, S., Kumar, A.: Low cost smart home automation system using smart phone. In: 2019 IEEE R10 Humanitarian Technology Conference (R10-HTC)(47129), Depok, West Java, Indonesia, 2019, pp. 120–125.https://doi.org/10.1109/R10-HTC47129.2019.9042467
3. Kotapalle, G.R., Kotni, S.: Security using image processing and deep convolutional neural networks. In: 2018 IEEE International Conference on Innovative Research and Development (ICIRD), 11–12 May 2018, Bangkok Thailand
4. Rammohana Reddy, E., Sankara, K.: Internet of things based home automation control system using Raspberry Pi. Int. J. Sci. Res. Comp. Sci. Eng. Inf. Technol. 2018 IJSRCSEIT **3**(4) (2018). ISSN: 2456-3307
5. Pravin Kumar, S., Balachander, B.: Fast and cost efficient face detection system with CNN using Raspberry PI. Int. J. Eng. Adv. Technol. (IJEAT) **8**(6). ISSN: 2249-8958, August, 2019
6. Xenya, M.C., Kwayie, C., Quist-Aphesti, K.: Intruder detection with alert using cloud based convolutional neural network and raspberry pi. In: 2019 International Conference on Computing, Computational Modelling and Applications
7. Srinivasan, S., Muthubalaji, S., Devadasu, G.: An improved h-bridge multi level inverter topology with minimal switches for harmonic reduction using artificial bee colony algorithm. Int. J. Adv. Sci. Technol. **29**(5), 5031–5040 (2020)

8. Rashid, E., Ansari, M.D., Gunjan, V.K., Ahmed, M.: Improvement in extended object tracking with the vision-based algorithm. In: Studies in Computational Intelligence, vol. 885, pp. 237–245 (2020)
9. Koppula, V.K., Sai Soumya, D., Merugu, S.: Nurse alarming device for bedridden patients using hand gesture recognition system. Lecture Notes in Electrical Engineering, vol. 643, pp. 377–385 (2020)
10. Prathyusha, C., Vijaya Shanthshanth, M., Harish, V.: Validation of solar PV-wind hybrid system with incremental conductance algorithm. J. Adv. Res. Dyn. Control Syst. **12**(4), 1168–1179 (2020)
11. Merugu, S., Jain, K., Mittal, A., Raman, B.: Sub-scene target detection and recognition using deep learning convolution neural networks. Lecture Notes in Electrical Engineering, vol. 601, pp. 1082–1101 (2020)
12. Vyshnavi, B., Srinivasa Rao, G.: Back-to-back switch connected NPC multilevel inverter FED in drive. J. Adv. Res. Dyn. Control Syst. **12**(4), 787–795 (2020)
13. Praveen Kumar, K., Amulya, B., Venkateshwar Rao, B.: Compact multiband printed antenna design and analysis. Int. J. Control Autom. **13**(3), 58–63 (2020)

Chapter 10
Image Security Based on Rotational Visual Cryptography

Ragipati Karthik, Veluthurla Mahammad Mobin Baji, Mittapalli Pavan Kumar, Shiza Arman Anjum, and Merugu Suresh

Introduction

Cryptographic procedure is applied on visual data much the same as photograph, notes, and print of text in VC. Distinct VC plots are discussed in this paper. Visual data are split into n number of shares or offers in the encoding process in which minimum of k offers is utilized for decoding of the encoded secret information [1]. Transmission of figures, number of offers that are encrypted and improvement of pixel quality are given more importance. If extension of pixel is less, then it leads to small offer size. When transporting various offers or mysteries, it takes less effort for encryption of image data into equivalent offer size. Programmer is required in processing of crucial shares is given high priority in the channel for security concerns [2, 3].

Related Work

Visual Cryptography

The first VC procedure is performed to create shares of the binary picture is developed by Naor and Shamir [1] which produces share 1 and share 2 offers of the given binary image. If by any chance, the color of the pixel is white than two lines are combined to

R. Karthik (✉) · V. M. M. Baji · M. P. Kumar · S. A. Anjum
Department of Electronics and Computer Engineering, Koneru Lakshmaiah Education Foundation, Vaddeswaram, Andhra Pradesh, India
e-mail: karthik39r@kluniversity.in

M. Suresh
Department of ECE, CMR College of Engineering and Technology, Kandlakoya, India

V. Garcia Diaz and G. J. Rincón Aponte (eds.), *Confidential Computing*,
Advanced Technologies and Societal Change,
https://doi.org/10.1007/978-981-19-3045-4_10

form share 1 and 2 [1, 4]. Correspondingly, if dark is the color of the pixel, then any one of two lines is chosen to generate share 1 and 2. Every share pixel *p* is encrypted into two dark and white pixels in which any one offer can't provide the information about the secret data either it is dark or white [5].

Color Visual Cryptography Scheme

Van Tilburg and Verheul developed the scheme structure at the beginning [6]. Curves are utilized to develop a color VC conspires for the colored pictures. If we consider *c* color VC, *c* color blocks are created with the help of sub-pixel which in turn changed from *m* pixels of one pixel plane. Pixel color of one pixel depends upon the inter-linkage of stacked pixels. This scheme will be having *c* colors and the pixel size *m* [7].

Previously, *n* shares are mandatory for the recreation of the image. The creators extended the procedure by using *k* minimum shares out of *n* shares, where each and every share is distributed among the clients. For the recreation of the image, we require minimum of *k* number of shares. If the number of clients containing shares is less than *k*, then the secret information of the picture is not revealed.

The inventors of VC had used another procedure known as anti-phishing technique using RSA methodology to provide security from phishing activities [2] where the original image is split into two shares, where client holds one of the shares, and employee holds another share. With the help of this process, security of the information is increased by utilizing VC procedure; RSA procedure gives ALGamal calculation [3].

The input picture is separated into ten squares which in turn converted to binary numbers by using VC procedure. By taking 64 bit as the size, the squares are separated into odd and even squares in which RSA is performed on even squares, and ALGamal calculation is performed on odd squares. Converting of squares to shares is a gained by using two-pixel encryption [4, 5].

Halftone Image Creation

Halftone image is created by performing Floyd–Steinberg procedure on the original picture [6]. By using this procedure, we can easily perform halftones on the color image. At first, original color image of any document type is input from the device. The pixels of the created halftone cannot be expanded anymore.

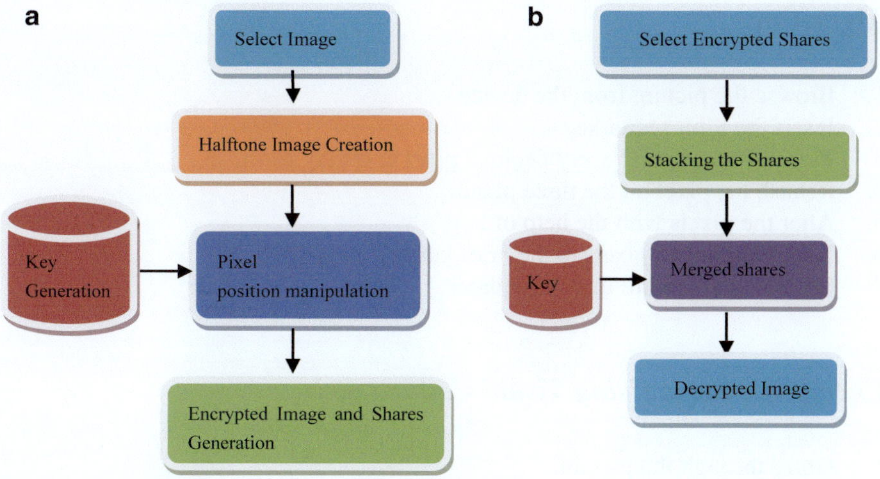

Fig. 10.1 **a** Encryption procedure **b** Decryption procedure

Share Generation

After the creation of halftones, then shares of the image can be created with the help of *k–n* sharing algorithm [7]. Here, ten number of shares are generated from the halftone image in VC and 6 shares as the minimum number of shares for the recreation of the original picture. If the shares taken for decryption procedure are less than 6, then the original picture is not obtained. The shares that are generated from halftone are having similar size, shape, pixel, on same level, and considerable images [8, 9] (Fig. 10.1).

Reconstruction

For the recreation of the original image, the shares that are generated by the help of the above procedure are mandatory. The minimum shares, i.e., *k* shares that are necessary for the creation of the picture are 6. These six shares are now stacked one by one on top of each other for creating the halftone picture. While stacking, the shares should be rotated [10]. The recreated halftone is now used to produce the original picture with the help of the secret key [11].

Algorithm

1. Browse the picture from the device.
2. Insert the 8-bit secret key.
3. Transform each and every decimal pixel to binary form.
4. Rehash the pixel on the three planes.
5. Alter the pixels with the help of key.
6. Now transfer the pixels to the brief lattice.
7. Separate the picture into ten squares.

Encryption Steps Using Visual Cryptography

1. Graze through the picture.
2. Acquire an element.
3. Split up the elements in red color, blue color, green color parts.
4. Convert this image to gray scale.
5. Generate shares from the grayscale image.
6. Produce encrypted image using key.
7. Save the encrypted image in hard disk.

Least Significant Bit Hiding Algorithm

Step 1 Select one pixel of input image randomly, divide the image in three shares [12]. Suppress one bit or section of unknown message in every division of element in least symbolic portions. Now, put the images with latest assess and reserve [13].

Step 2 Carving the figure using VC, the figure is break down to different allotments. The allotments produced to disclose no actual date regarding the primary unknown image.

Step 3 Stockpile the figure at the grantee side using VC.

(1) Later granting the allotments go through all the elements and place that in the adept slide.
(2) Then, show the adept figure on the serene.
(3) Redo the procedure for every allotment.

Step 4 Decoding using stenography.

(1) Repeat the procedure for every one of the pixels.
(2) Retrieved bits to be reversed.
(3) Put together the characters of string.
(4) Display the retrieved message.

Visual Cryptography

Engraving VC by engraving narrative approach achieves VC [14]. Built upon the azure-sound waver assumptions, technique holds up the cluster and void computations to cipher the classified double-edged pictures into halftoned allotments that give important visual data [15].

Image

An figure can be conveyed as a two-dimensional function $f(p, q)$ where p and q are referred as the coordinates, and the image intensity is the amplitude of 'f' at that distinct place.

Gray Scale Image

Intensity of the classified image is $I(x, y)$ at the point (x, y) on plane. It is bounded by a rectangle and takes non-negative values on the image [16].

Color Image

Color image can be expressed as three functions, $R(x, y, l, e, m)$ for red color, $G(x, y, l, e, m)$ for green color, and $B(x, y, l, e, m)$ for blue color. The image is continuous in means of coordinates and also in amplitude. Converting image to digital bits involves coordinates and amplitude of classified image to be digitalized. Sampling is known as digitalizing the arranged assess [17].

Coordinate Convention

Convert the input figure $f(x, y)$ into N columns and M rows. So that the image is $M \times N$ size. The coordinates values are (x, y, l, e, m) distinct. For the notational clarity and convenience, discrete coordinates utilize integer values. And the notation (0, 1) defines second sample along first row.

Table 10.1 Image formats

Format name	Description	Extension
JPEG	Joint photographic experts group	.jpeg, .jpg
PNG	Portable network graphics	.png
BMP	Windows bitmap	.bmp
GIF	Graphics interchange format	.gif
TIFF	Tag image file format	.ti, .tif
XWD	X window dump	.xwd

Reading Image

In MATLAB, imread function is used to import, and the syntax is

Imread('file name')

In Table 10.1, image format types are included in which a secret image of the above one of the format can be used to import the image in the MATLAB environment.

Data Classes

The allotment assess is dissimilar to be numeral values in MATLAB. We focus on both class unit 8 and class unit 16 that consist of the initial information classes. Many tasks encourage all the information classes.

MATLAB Environment for Visual Cryptography

The MATLAB editor can be used as both graphical MATLAB debugger and a text editor. Image representation can be done by using rectangular array of pixels (values). Every pixel defines some property of measured scene in a limited area. Here, property represents the measure of brightness of secret image which is filtered by three filters; those are red, blue, and green filters expressed as three values or single value as average brightness. The values are expressed by 8-bit integers in a range containing 256 brightness levels. Image resolution can be expressed as number of brightness values and number of pixels. MATLAD can read all the image formats mentioned in Table 10.1.

Image loading will be done in working memory with help of command which is expressed in Eq. (10.1).

$$f = \text{imread}(\text{'image name'}); \tag{10.1}$$

Table 10.2 Conversion of images

Function	Input type	Output type
im2uint8	Logical, uint8, uint16, double	Uint8
im2uint16	Logical, uint8, uint16, double	Uint16
mat2gray	Double	Double in range [0, 1]
im2double	Logical, uint8, uint16, double	Double
im2bw	Uint8, uint16, double	Logical

Image displaying and image saving can be done by using Eqs. (10.2) and (10.3)

$$\text{imshow}(f, G) \tag{10.2}$$

$$\text{imwrite}(\text{array name},\ \text{'file name'}) \tag{10.3}$$

Gray-level images are represented in pixels in every channel that can be expressed as 8-bit integers.

MATLAB functions shown in Table 10.2 are used to change images from one form to another form. MATLAB has many methods of obtaining the image pixels or array elements of image. The function in Table 10.2 can be convened as

$$B = \text{uint8}(A) \tag{10.4}$$

In the above Eq. (10.4), A represents the array of elements of the image. Histogram is helpful in change of grayscale image to acquire different forms. Coordinate manipulation is used to distort the classified image, and the high-pass edge detection is expressed as local and significant changes in real time in the intensity of an image.

Results and Discussion

Select the secret image that is to be encrypted which is in Fig. 10.2. The selected secret image is now converted to halftone image (grayscale image). The created halftone image is shown in Fig. 10.3. Shares and encrypted image is generated by using secret key.

Figures 10.4 and 10.5 represent the shares generated and encrypted image by using 6-bit key. Ten shares are generated in shares generation for encryption technique.

For decryption process, select encrypted image and shares one by one. Now, by using selected shares, stack each share one by one and merge encrypted image. Decryption can be achieved using a secret key of the encrypted image. After the decryption process is completed, the secret image is generated.

Fig. 10.2 Image to be encrypted

Fig. 10.3 Halftone image creation

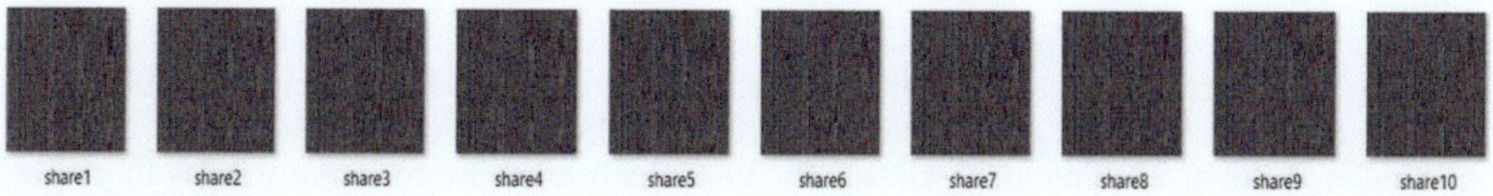

Fig. 10.4 Shares generation

Fig. 10.5 Encrypted image

Conclusion

Encryption and decryption of the image can be done using cryptographic technique by using a secret key. The encryption is based on many parameters, and in the same way, basing on the previous parameters which are taken during encryption is utilized for the decryption process. Encryption can be done by share generation, and decoding can be done by stack piling of encrypted offers one by one.

References

1. Naor, M., Shamir, A.: Visual cryptography. Advances in cryptology. In: Eurocrypt'94 Proceeding LNCS, vol. 950, pp. 1–12 (1995)
2. Chauhan, R., et al.: Estimation of software quality using object oriented design metrics. Int. J. Innov. Res. Comput. Commun. Eng. **2**(1), 2581–2586 (2014)

3. Mohamed, R.M., Kadhim, A.: Visual cryptography for image depend on RSA & AlGamal algorithms. In: 2016 Al-Sadeq International Conference on Multidisciplinary in IT and Communication Science and Application (AIC-MITCSA)—IRAQ 9, 10 May
4. Sindhu Parkavi, S., Sharon, I., Gowri, S.: Visual cryptography for color images to provide confidentiality using embedded system. Glob. J. Pure Appl. Math. **13**(6), 2555–2561 (2017). ISSN 0973-1768
5. Rathore, S., Verma, N.: A secure secret shares by novel visual cryptography using bit rotation and blowfish algorithm. In: IJESRT, May, 2014
6. Singh, O., Bommagani, G., Ravula, S.R., Gun-Jan, V.K.: Pattern based gender classification. In: IJARCSSE, vol. 3, pp. 888–895, October 2013
7. Saturwar, J., Chaudhari, D.N.: Secure visual secret sharing scheme for color images using visual cryptography and digital watermarking. In: IEEE 2018
8. Suma, D., Raviraja Holla, M.: Pipelined parallel rotational visual cryptography (PPRVC). In: International Conference on Communication and Signal Processing, April 4–6, 2019, India
9. Balakrishna, S., Thirumaran, M., Solanki, V.K., Gunjan, V.K.: Performance analysis of linked stream big data processing mechanisms for unifying IoT smart data. In: Proceedings of International Conference on Intelligent Computing and Communication Technologies (ICICCT), 2019, Hyderabad, India. Springer, pp. 680–689
10. Kashyap, A., et al.: Computational and clinical approach in lung cancer detection and analysis. Procedia Comput. Sci. **89**, 528–533 (2016)
11. Lee, K.-H., Chiu, P.-L.: Threshold visual cryptography schemes with tagged shares. Ministry of Science and Technology of Taiwan. https://doi.org/10.1109/ACCESS.2020.3000308
12. Yang, C.-N.: k out of n region-based progressive visual cryptography. In: IEEE Transactions on Circuits and Systems for Video Technology. https://doi.org/10.1109/TCSVT.2017.2771255
13. Bala Dastagiri, N., Hari Kishore, K., Gunjan, V.K., Fahimuddin, S.: Design of a low-power low-kickback-noise latched dynamic comparator for cardiac implantable medical device applications. In: Satapathy, S., Bhateja, V., Chowdary, P., Chakravarthy, V., Anguera, J. (eds.) Proceedings of 2nd International Conference on Micro-Electronics, Electromagnetics and Telecommunications. Lecture Notes in Electrical Engineering, vol. 434. Springer, Singapore (2018). https://doi.org/10.1007/978-981-10-4280-5_67
14. Krishnaveni, K., Venkata Ratnam, K., Prathyusha, G., Gopi Krishna, P.: Development of real time environment monitoring system using with MSP430. Int. J. Eng. Technol. (UAE) **7**(2), 72–76 (2018)
15. Shaik, A.S., Usha, S.: Sensor based garbage disposal system. Int. J. Inno. Technol. Exp. Eng. **8**(452), 164–167 (2019)
16. Singh, R., Chauhan, R., Gunjan, V. K., Singh, P.: Implementation of elliptic curve cryptography for audio based application. Int. J. Eng. Res. Technol. (IJERT) **3**(1), 2210–2214 (2014)
17. Sri Laxmi, P., Kumar Sanjiv, B., Khare, V.: Design and comparison of different 4:2 compressors based on 180 nm technology. Int. J. Inno. Technol. Exp. Eng. **8**(452), 147–150 (2019)

Chapter 11
Development of Safety Monitoring for an IOT-Enabled Smart Environment

A. Harshitha, Manikanta Uma Srinivas, M. Eswar Sai, Krishnaveni Kommuri, and P. Gopi Krishna

Introduction

Reconnaissance cameras are available in a wide extent of styles and features, and they are an average fragment in a security system. Web security structures or IP cameras use the Internet by frameworks organization to send and get data. They are truly easy to present and interface with your structure, and you can see live camera deals with at whatever point with free adaptable applications for cutting edge cells and tablets. IP cameras can be presented in essentially any region and can screen within similarly as the outside of an office. Cameras can in like manner be used to alert security/protection authorities to respond to questionable activities or individuals. Video perception incorporates the exhibit of watching a scene or scenes and looking for express practices that are unseemly or that may show the turn of events or presence of foolish direct. Essential businesses of video observation fuse watching individuals all in all at the entry to games, public transportation (train stages, air terminals, etc.), and around the edge of secure workplaces, especially those that are honestly restricted by network spaces. The endeavor by then widens to the library of OpenCV. It is a library of Python attaches expected to handle PC vision issues. OpenCV-Python uses NumPy, which is an extraordinarily progressed library for numerical assignments with a MATLAB-style etymological construction. All the OpenCV bunch structures are changed over to and from NumPy displays.

A. Harshitha · M. U. Srinivas · M. E. Sai · K. Kommuri (✉) · P. G. Krishna
Department of ECM, Koneru Lakshmaiah Education Foundation, Vaddeswaram, Andhra Pradesh, India
e-mail: krishnavenikommuri@kluniversity.in

V. Garcia Diaz and G. J. Rincón Aponte (eds.), *Confidential Computing*, Advanced Technologies and Societal Change,
https://doi.org/10.1007/978-981-19-3045-4_11

Video Processing

Computer vision may be a mammot 3.1 Ultrasonic detector:h piece of the data science/AI space currently, and once more, computer vision engineers ought to manage recordings. Here, we tend to hope to uncover understanding into video preparing—victimization Python, clearly. This could be evident for some, but all things thought about video real time is under no circumstances a standardized cycle, nonetheless a distinct one. That means, anytime we tend to manage recordings, we tend to area unit managing the succession of edges themselves. Every casing is simply an image, which can be addressed as a $m \times n$ exhibit of pixels, where (m, n) is image size. Each element is also addressed as shading power; contingent upon that shading model, we tend to area unit utilizing (gray scale, RGB, or maybe multispectral. The elemental video handling equipment for Python—OpenCV. OpenCV is associate degree ASCII text file library which supplies USA the gadgets to perform primarily such a picture and video handling. OpenCV is written in C++, and its essential interface is in C++. cv2.waitKey() may be a necessary structure block for OpenCV video handling. Stand by secret is a technique that shows the casing for indicated milliseconds. The "0xFF == Ord ('q')" within the "if" proclamation is a rare punctuation to offer the "while" circle break, by a console key squeeze occasion. cap. Delivery () and cv2.destroybAllWindows () area unit the methods to shut video documents or the catching gismo, and extinguish the window that was created by the show technique. That is, presently, we tend to notice a way to trot out recordings in Python, and that we notice a way to get to every casing of a video. From here, we will add any style of image handling within this cycle; it can be either object identification, or human posture assessment model, or both. The additional jump into OpenCV for video making ready is up to per user. There square measures a bunch of useful instruments which will meet any of your necessity, from the smallest amount hard to please assignment (resize, crop, and shading/splendor change), passing by image separation to triangulation.

OpenCV and NumPy

OpenCV

OpenCV is a huge open-source library for computer vision, machine learning, and image processing. OpenCV supports a wide variety of programming languages like Python, C++, and Java. It can process images and videos to identify objects, faces, or even the handwriting of a human. When it is integrated with various libraries, such as NumPy which is a highly optimized library for numerical operations, then the number of weapons increases in your Arsenal, i.e., whatever operations one can do in NumPy can be combined with OpenCV. Python may be a library of Python attaches planned to cope with laptop vision problems. Python is associate degree extensively useful programming language started by Guido van Rossum that complete up being uncommonly normal quickly, essentially by virtue of its straightforwardness and code clarity. It engages the designer to convey musings in less lines of code while not decreasing heaviness.

Stood out from vernaculars like C/C++, Python is a lot of slow. In lightweight of everything, Python will be viably disentangled up with C/C++ that grants United States of America build to create computationally raised code in C/C++ and make Python covers which will be used as Python modules. This provides United States of America two positive conditions: First, the code is just about as quick because the principal C/C++ code (since it is the real C++ code operating in establishment), and second, it easier to code in Python than C/C++. OpenCV-Python may be a Python covering for the principal OpenCV C++ execution. OpenCV-Python uses NumPy that is associate degree unbelievably improved library for numerical exercises with a MATLAB-style history style. All the OpenCV bunch structures area unit modified over to and from NumPy shows. This in like manner makes it easier to consolidate with numerous libraries that use NumPy, for instance, SciPy and Matplotlib.

NumPy

NumPy is a library for the Python programming language, adding support for enormous, multidimensional clusters and frameworks, alongside a huge assortment of undeniable-level numerical capacities to work on these arrays. The predecessor of NumPy, numeric, was initially made by Jim Hugunin with commitments from a few different designers [1]. In 2005, Travis Oliphant made NumPy by joining provisions of the contending Numarray into numeric, with broad alterations. NumPy is open-source programming and has numerous supporters.

NumPy is a broadly useful exhibit preparing bundle. It gives a superior multidimensional exhibit article and instruments for working with these clusters.

It is the principal bundle for logical figuring with Python. It contains different provisions including these significant ones:

(1) A incredible *N*-dimensional exhibit object
(2) Sophisticated (telecom) capacities
(3) Tools for incorporating C/C++ and Fortran code
(4) Useful straight variable-based math, Fourier change, and arbitrary number abilities.

Other than its undeniable logical uses, NumPy can likewise be utilized as a productive multidimensional compartment of conventional information.

Discretionary information types can be characterized utilizing NumPy which permits NumPy to flawlessly and expediently coordinate with a wide assortment of datasets.

Related Work

In this paper, various weakening models like ITU-R, RH, surface-to-air missile, and Mapfumo area unit contemplated, and also, the results area unit differentiated and calculable qualities and poor right down to decide the affordable model for one [2]. During this paper, we tend to review the potential attacks with reference to Cisco-Seven layer model. Also, preparing force, energy verge of collapse, and cutoff limits of very little enrolling gadgets have essentially overhauled, whereas their sizes have diminished whole [3]. These sensible conditions create them to be force in toward the new mechanical rise within the field of ineluctable registering. Middleware assumes a basic half in building the ineluctable applications. The inevitable gadgets act obsessed on the setting of the circumstance, that is, they are doing their activities as indicated by the climate of the applying. They answer the circumstances sagaciously as they will take their own selections obsessed on the setting created for that individual application [4]. Here, the protocol execution problems come about for the foremost half due to blunders in transmission and handoffs. The paper presents an intensive study of various methodologies with reference to the development in protocol execution in remote organizations. It sums up the various planned procedures and presents the benefits and downsides of these methodologies [5]. Another channel that might satisfy an oversized portion of the wants of a current when administrative official was created during this half. Hearty assessment area unit procedures are created for a few regular problems, as an example, assessing space, scale [1].

A basic piece of home robotization is to possess every home provided with checking frameworks to computerize homes. Here comes the requirement of a stripped-down effort Wi-Fi authorized regulator to assemble info and distribute it to employee. Hub MCU is one such unimaginable alternative. Utilizing this framework home natural conditions is checked, and also, the info is sent through MQTT convention, and it is distributed through employee to customers [6]. This paper indicates concerning the checking ecological boundaries systematically to conjecture the climate expectation. These days, prognosis assumes an essential half for residents living within the city district regions. Typically, the boundaries of the climate forecast amendment to varied fields and regions that hip to the horticultures and travelers. MSP430 It is employed for manufacturing plant management and automation applications [7]. This paper proposes associate degree approach to assemble a financially savvy ecological checking framework utilizing NodeMCU. The environmental factors during which individuals, creatures, or plants live are termed surroundings. Recognition of climate boundaries like temperature, stickiness, greenhouse emission et al., these variables assume a large half because it is squarely connected to medical issue of people within the climate [8]. DCNN with another double characterization layer that is reinvigorated on the online. This cycle empowers net based mostly following by ill the ROI windows discretionarily tested near to the past ROI state. It helps accomplishing constant vehicle following abundant below impediment and dynamic foundation conditions [9]. During this paper, they are talking a couple of general survey on wearable and advantageous sturdy devices for externally weakened people. There area unit various procedures, but all told of gadgets; they cannot acknowledge the speed of the vehicle and partition between the individual and vehicle. During this association, we will provide a theoretical system on new IoT-authorized coordinated convenience to assist externally debilitated people. The planned gismo will notice the speed and distance of a vehicle that is coming back toward the shopper [10].

The idea of recognizing a moving item from a comparative foundation is called as covering. During this audit study, numerous methods for moving factor identifying proof with some highlights planned by totally different scientists were examined [11]. Article affirmation in innovative photos is finished exploitation grammar. This paper advocates a completely unique system for removing information concerning the states of assorted things and components during a high-level image and for seeing articles employing a neural association [12]. Downy principle may be a thanks to affect reckoning obsessed on "levels of truth" rather than the quality factor "valid or bogus." Numerous sorts of nonlinear channels like expanded Kalman channel, cross-breed broadened Kalman channel, and firefly 0.5 breed broadened Kalman sift are tried to figure through the perfect arrangement. Goof based mostly feathery reasoning has been the rationale of calibration the methodology [13]. TensorFlow bundles along various AI and important learning models and counts and makes them accommodating by technique for a run of the mill likeness. During this paper, the count for moving article in static institution based mostly circumstance has been planned. This assignment is by and huge performed by institution allowance procedure and factor finder [14]. TensorFlow is a begin to complete ASCII text file AI stage for everyone. During this paper, they gift article acknowledgment utilizing Kera's Library with backend Tensor stream to remain off from troubles in acknowledgment of things in photos; the profound neural organizations notably Tensor stream below Kera's Library is employed [15].

We can show that the planned cross-selection model accomplished a better f-extent of 76.7% appeared otherwise in respect to that of different existing when models, as an example, the closest neighborhood algorithmic model and spatial-dramatic weighted moving typical model [16]. Presently, on a daily basis, Crowd investigation become associate degree arising field in exploration. The safety issue is attended by ways for mechanized examination of cluster exercises utilizing management video. The principle components of Crowd Analysis, viz., assessment of size of the cluster, social following, and development assessment area unit talked concerning during this paper [17]. One important advance is 3D scenes age from second photos. One among the tough and hard errands is age of Depth Map from Single read Image. Profound learning used here for seeing from image and also the significance map created [18]. The system connected with object acknowledgment fuses division, plan, and collection of articles which will be associated in numerous designing. During this paper, a completely unique portrayal framework known as wealth based mostly image classification (AIC) is planned employing a riffle neural framework [19]. This paper objections to propose a groundbreaking atom channel subject to changed timber wolf optimizer (MGWO) which can beat the impoverishment of the model issue within the normal particle channel.

Radio frequency identification (RFID) may be a remote development used for following a reputation related to a factor and remarkably basic cognitive process it. Associate degree alpha examination of the created model of the overhauled shows gathering device is performed, and also, the results affirm its unimaginable introduction. The examination of Web site Diversity Gain Model dependent on the aware precipitation information has been finished. The radio waves multiplying through the planet surroundings are tightened on account of the presence of measuring device particles, as an example, water smolder, storm drops, and also the ice particles that absorb and scatter the radio waves and, on these lines, degenerate the introduction of the microwave interface. These assumption models have the key information, the geologic zone, the repeat of the sign used, and also the proportion of precipitation at that zone. By applying these models at totally different geographical territories, enervating of the association for level of the time and with numerous frequencies within the Unq band vary, i.e., 10.99 rate to fourteen.2 rate area unit expected.

Radio wires performing at gigahertz frequencies area unit being planned and dead for satellite correspondences. Yet, the microwave signals area unit encountering loss of sign strength once meddled with totally different layers of the surroundings, precipitation, mists so on pelter drop circulation supplanted this tough work with disentangled investigation for a selected scene. During this examination, distinctive existing bunch based mostly, EM grouping, and game arrange based mostly irregularity acknowledgment systems in video observation area unit talked concerning. The new methodologies just like the mixture of convolution neural network (CNN) and repeated neural network (RNN) and course profound learning area unit the powerful calculations for immense datasets.

Safety Monitoring for Smart Environment System Overview

After associate degree audit of the connected works, turning out next is the principle goals that we have set for the projected framework: (a) endlessly screen the climate for boundaries, for instance, separation of the centered on object, soaker water recognition, object location (b) ultrasonic detector and soaker detector module need to be related to the Raspberry Pi to screen the IoT climate (c) camera will interface by utilizing simply OpenCV strategy or by we {willlwe are able to} interface with Raspberry Pi through OpenCV technique (d) And GSM is to boot utilized for causing prepared message to the consumer presumably; it is running or any article identification to alarm the consumer that the gismo will come back to update state

Ultrasonic Detector

Supersonic sensor is associate degree device that measures the space of a target object by emitting supersonic sound waves associate degreed converts the mirrored sound into an electrical signal. Supersonic waves travel quicker than the speed of loud sound (i.e., the sound that humans will hear). Supersonic sensors have two main components: the transmitter (which emits the sound victimization electricity crystals) and also the receiver (which encounters the sound once it's cosmopolitan to and from the target). Ultrasonic sensors are often used for several applications, together with precise detection of objects and contactless observance of fill levels. Supersonic detectors add abundant identical means as measuring device and measuring system. They produce high-return sound waves and assess the reverberation that is gotten back by the sensor.

Ultrasonic transducers and ultrasonic sensors are gadgets that create or sense ultrasound energy. They can be partitioned into three general classes: transmitters, beneficiaries, and handsets. Transmitters convert electrical signs into ultrasound, beneficiaries, convert ultrasound into electrical signs, and handsets can both send and get ultrasound.

Ultrasonic transducers convert AC into ultrasound, just as the converse. Ultrasonics, normally, alludes to piezoelectric transducers or capacitive transducers. Piezoelectric gems change size and shape when a voltage is applied; AC voltage causes them to sway at a similar recurrence and produce ultrasonic sound. Capacitive transducers utilize electrostatic fields between a conductive stomach and a support plate.

The bar example of a transducer can be controlled by the dynamic transducer region and shape, the ultrasound frequency, and the sound speed of the spread medium. The outlines show the sound fields of an unfocused and a centering ultrasonic transducer in water, clearly at varying energy levels.

Since piezoelectric materials create a voltage when power is applied to them, they can likewise fill in as ultrasonic identifiers. A few frameworks utilize separate transmitters and beneficiaries, while others consolidate the two capacities into a solitary piezoelectric handset.

Ultrasound transmitters can likewise utilize non-piezoelectric standards like magnetostriction. Materials with this property change size marginally when presented to an attractive field and make down to earth transducers.

A capacitor ("condenser") amplifier has a slender stomach that reacts to ultrasound waves. Changes in the electric field between the stomach and a firmly dispersed sponsorship plate convert sound signs to electric flows, which can be intensified.

The stomach (or film) guideline is likewise utilized in the moderately new miniature machined ultrasonic transducers (MUTs). These gadgets are manufactured utilizing silicon miniature machining innovation (MEMS innovation), which is especially valuable for the creation of transducer exhibits. The vibration of the stomach might be estimated or initiated electronically utilizing the capacitance between the stomach and a firmly dispersed sponsorship plate (CMUT) or by adding a slim layer of piezoelectric material on stomach (PMUT). Then again, late exploration showed that the vibration of the stomach might be estimated by a minuscule optical ring resonator coordinated inside the stomach (OMUS)

Rain Detector

A rain sensor is one reasonably shift device that is employed to observe the precipitation. It works sort of a switch, and also, the working rule of this detector is whenever there is rain, the switch is commonly closed. **Rain detector Module**: Essentially, the board includes nickel-coated lines, and it works on the resistance principle. This detector module permits to determine wetness through analog output pins, and it offers a digital output, whereas wetness threshold surpasses.

In a world loaded with robotization, everybody needs to remain savvy and useful. What is more, water collecting can likewise be accomplished effectively with the assistance of computerization. Here, in this article, we will examine the Rain Alarm project which has an extraordinary importance in collecting water. The circuit assists with distinguishing the downpour by giving a caution, while you are sitting inside the home, so you can make a fundamental move. It thoroughly helps in staying away from the superfluous cerebral pain of keeping an eye outside your window in the stormy season to see when it will rain. To set aside your time cash and exertion, how about we start with the most common way of making a basic and financially savvy downpour alert circuit in the least demanding manner.

This circuit enables you to manage your outside settings when it rains. You can close your windows or get ready for rain water harvesting after hearing the alarm. With some small modification, the circuit can also be used as water level indicator, agriculture field irrigation system, etc. The biggest advantage of this circuit is that it is simple, reliable, and consumes less power.

When there is no downpour, the obstruction between the wires of downpour sensor will be extremely high, and there will be no conduction between the wires on sensor.

At the point, when the downpour drop falls on the sensor, it will frame a conductive way between the wires, and the opposition between wires will diminishes. Now, the wires on sensor board will begin leading current.

Essentially, this sensor is filling in as a switch, so at whatever point water is available on the sensor, it permit the momentum to go through it (ON condition). What is more, on the off chance that there is no water, it ends the progression of momentum (OFF condition).

GSM

Worldwide framework versatile correspondences were created as a computerized framework utilizing time division completely different access (TDMA) procedure for correspondence reason. A GSM digitizes and lessens the data, at that time sends it down through a channel with two distinct surges of client info, every in its own specific time allotment. The computerized framework features a capability to convey sixty four kbps to a hundred and twenty Mbps of knowledge rates. The system is capable enough to instruct user via SMS from a selected signal to vary the condition of the house appliance per the user's desires and necessities. The second facet is that of security alert that is achieved in an exceedingly means that on the detection of intrusion, the system permits automatic generation of SMS therefore alerting the user against security risk.

The Global System for Mobile communications (GSM) is a standard created by the European Telecommunications Standards Institute (ETSI) to portray the conventions for second-age (2G) computerized cell networks utilized by cell phones like cell phones and tablets.

Fig 11.1 addresses the framework usage and activity. We have utilized Raspberry Pi board for interfacing all the sensors of ultrasonic and downpour sensor. OpenCV technique for reconnaissance camera. Triangulation technique is executed for camera position. Furthermore, GSM module is likewise associated with the Raspberry Pi for sending ready message to the client since it is sufficiently skilled to teach client through SMS. The subsequent viewpoint is that of security ready which is accomplished in a manner that on the identification of interruption, the framework permits programmed age of SMS along these lines cautioning the client against security hazard.

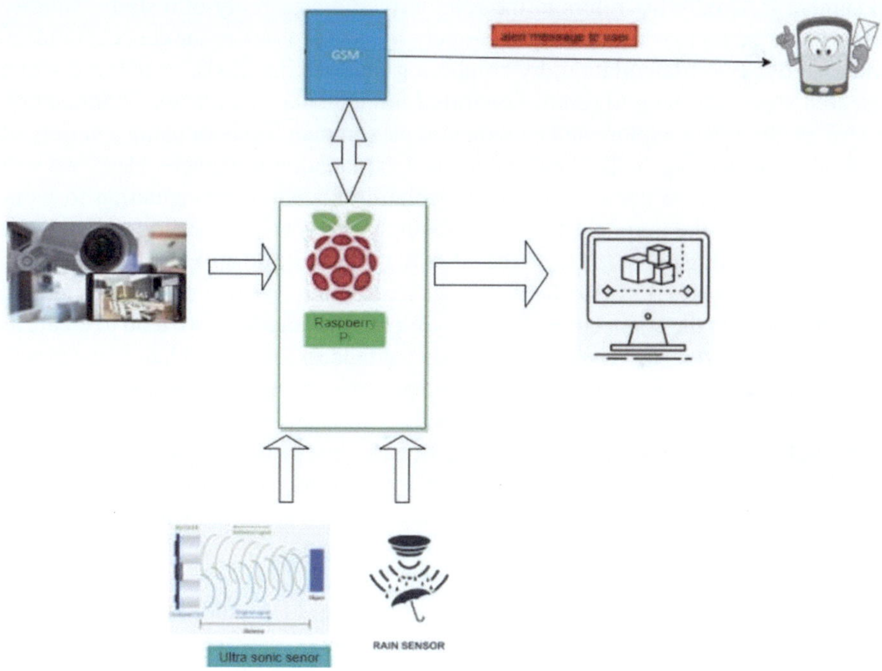

Fig. 11.1 Smart environment system overview arrangement

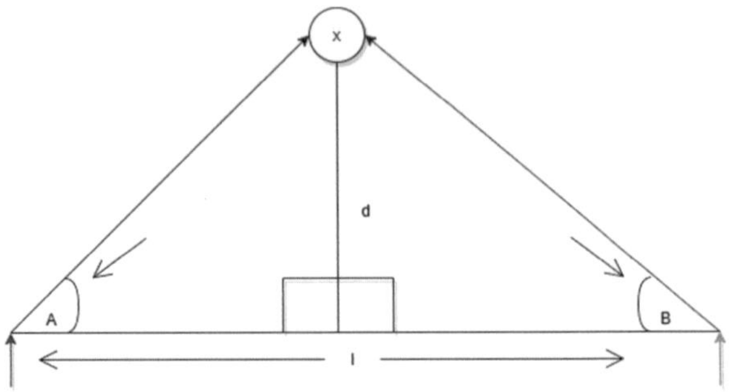

Fig. 11.2 Setting the cameras position by triangulation

Triangulation Method for Camera Arrangement

From Fig. 11.2 address left camera with associate degree angle (A) and right camera with purpose (B) and discovering purpose within the area addresses an object and expect that two cameras lies on a comparative plane. notice we discover} the purpose from object to camera associate degree, and that we find the purpose from object to camera B. Credibility refers to the reliability and credibility of a study; validity refers to the extent to which a study accurately reflects or assesses the concept or ideas under study. Two triangulation, by combining theories, methods, or observers in a research study, can help to remove distorted fundamentals that arise. Triangulation is also an attempt to explore and explain complex human behavior using a variety of methods in order to provide a more balanced explanation to readers. Here, we will use a beautiful range connected issue calculation known as triangulation to work out {the purpose the purpose} that in area simply an issue and creating a right point triangle by drawing a line (d) from the issue to the plane, and also, the total partition between 2 cameras is 'l'. In mathematics, we tend to perceive that diversion is akin to the converse over the connecting. Thus, to seek out the length (l), we tend to create as $l = d/\tan(A) + d/\tan(B)$. The essential issue we tend to found the chance to possess by them could be a purpose from camera and a degree from camera B. $\tan = 0/a$, $\tan(39) = 320/x$, $d = 320/\tan(39)$, $d = 395$.

From Fig. 11.3, it represents to a casing of a camera with pixels 640p; we are able to utilize any scope of pixels. So, the camera is all the way down to the sting, and our camera contains a field of vision which camera vision produces with sure purpose from those points; we have a tendency to area unit finding the separation consistent with the equation.

We need to orchestrate the two cameras as per Fig. 11.4. Additionally, check the cameras, wherever there square measure examination to every alternative or not and what is more, we want to urge three hatchets indisputably. By running the code, it shows the detachment of the zeroed in on article on the screen.

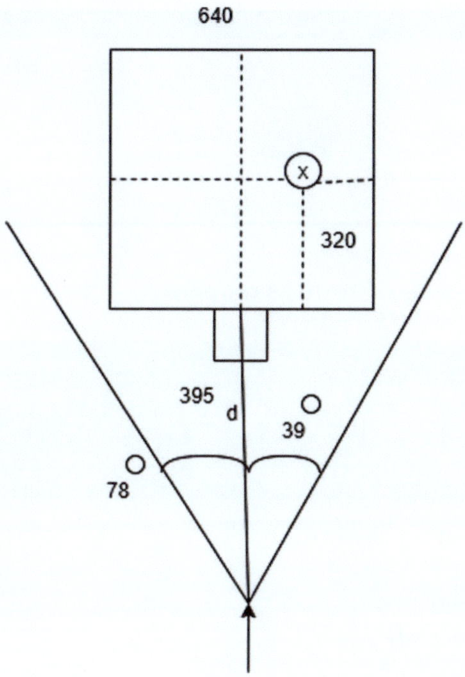

Fig. 11.3 Frame of the camera

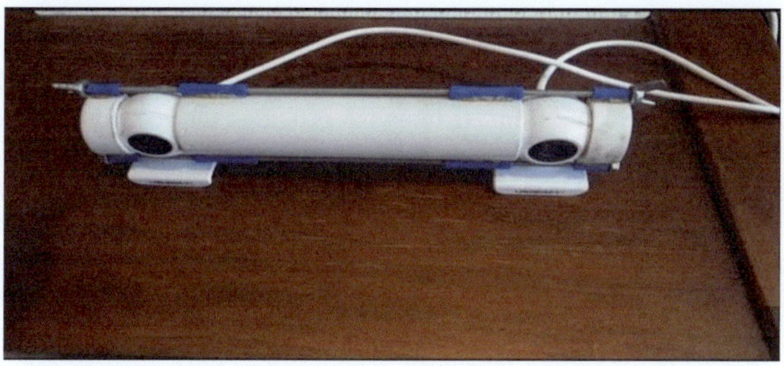

Fig. 11.4 Camera setup

Results and Discussions

The code runs adequately and showed the gap of the targeted on article and gave security to the widget to anticipate the update condition of the widget. Here is that the screen capture of the yield in Fig. 11.5 regarding video that had sent. An alarm messages to the shopper through GSM module. We will see a warning once it acknowledges the movement of a piece of writing or a private. Here, we will see

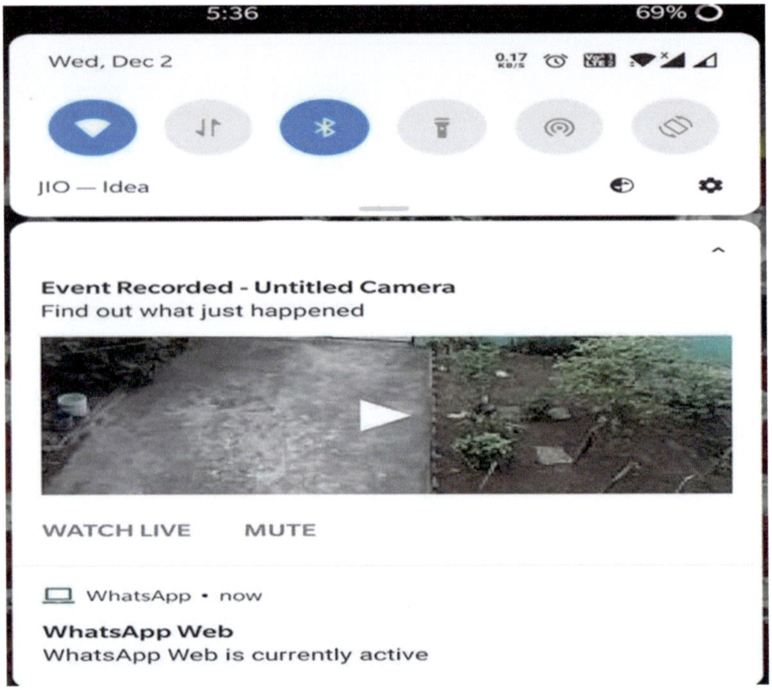

Fig. 11.5 Notification from camera

that notice was shipped off the tip shopper as sort of a video it begins giving the admonition at no matter purpose the article is known. Therefore, we will get the clear thought relating to the item. It likewise provides the live to observe what is happening.

From Fig. 11.6, it is clear yield for movement discovery. We are able to see the excellence from two footages. Also, we have a tendency to recognize the movement recognition in see Fig. 11.5 it known with the help of observation camera. Like these, we have a tendency to effectively finish the venture by utilizing the triangulation strategy. During this technique, we have a tendency to analyze the two occasions for reference on the off likelihood that there is no modification in each the examples; at that time, they clear their info and once more begins filtering. If they recognized something, whereas at a similar time different each the cases, then they consequently build the screen efforts and send the warning to the tip consumer within the higher than given image; we are able to see that associate item is passed before camera; therefore, it analyzes if there is any modification by utilizing the last case that is employed for the reference. At that time, it quickly sends the admonition message to the consumer.

Here, we can see that the any moment occurs in front of the camera. It will be tracked and mailed to the given user mail id so that the user can able to know who came into their surroundings.

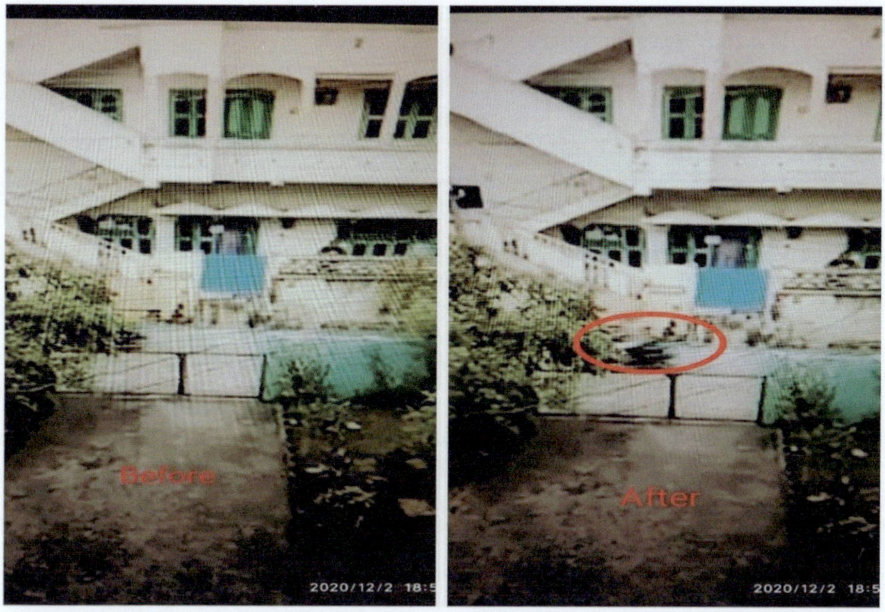

Fig. 11.6 Motion detection with the experimental setup

Many of the unwanted things are can be sent to mail by camera, but at first, we have to train it accordingly so that it can get clarity which of the things we have to send and the things which are not.

Conclusion

The venture was finished effectively, and the ideal codes were executed. The momentary point includes detail investigation of the video preparing dependent on triangulation for computer vision. Also, we have effectively added security by adding other two modules for the reconnaissance. An exertion was made to assemble greatest information about it from individuals who have accomplished exemplary work in the field. This was trailed by a usage of a similar utilizing standard informational collection. We have additionally intended to investigate more methods of computer vision in not so distant future. Drawn out objective is to have the option to actualize a method where the obscured pictures could be de-obscured with a degree of adequate precision. Investigated different existing approaches for de-obscuring strategies where an obscured picture is honed and smoothened to make it more comprehendible for human use.

References

1. Reddy, A.V.N., Phanikrishna, C.: Contour tracking-based knowledge extraction and object recognition using deep learning neural networks. In: Proceedings on 2016 2nd International Conference on Next Generation Computing Technologies, NGCT 2016, pp. 352–354 (2017)
2. Immadi, G., Venkata Narayana, M., Suraj, Y., Nara Simha Rao, N.M.V.L., Naveen Chowdary, P.S.V.S., Emmanuel Raju, M.: Estimation of effect of troposphere rain on radio link in tropical environment. ARPN J. Eng. Appl. Sci. **12**(17), 4960–4966 (2017)
3. Balakrishna, S., Thirumaran, M., Solanki, V.K., Gunjan, V.K.: Performance analysis of linked stream big data processing mechanisms for unifying IoT smart data. In: Proceedings of International Conference on Intelligent Computing and Communication Technologies (ICICCT), 2019, Hyderabad, India, pp. 680–689. Springer
4. Madhusudanan, J., Geetha, S., Prasanna Venkatesan, V., Vignesh, U., Iyappan, P.: hybrid aspect of context-aware middleware for pervasive smart environment: a review. Mobile Information Systems, 1–16. Article ID 6546501 (2018)
5. Kamble Rita, R., Rajarajeswari, P.: A review of various camouflage moving object detection techniques. J. Eng. Appl. Sci. **12**(Special issue 12), 9514–9519 (2017)
6. Bala Dastagiri, N., Hari Kishore, K., Gunjan, V.K., Fahimuddin, S.: Design of a low-power low-kickback-noise latched dynamic comparator for cardiac implantable medical device applications. In: Satapathy, S., Bhateja, V., Chowdary, P., Chakravarthy, V., Anguera, J. (eds.) Proceedings of 2nd International Conference on Micro-Electronics, Electromagnetics and Telecommunications. Lecture Notes in Electrical Engineering, vol. 434. Springer, Singapore (2018). https://doi.org/10.1007/978-981-10-4280-5_67
7. Das, R.P., Roja, G., Sampath Dakshina Murthy, A., Koteswarao Rao, S.: Global positioning system object tracking by applying fuzzy logic nonlinear techniques. J. Adv. Res. Dyn. Control Syst. 9, 725–735 (2017)
8. Chauhan, R., et al.: Estimation of software quality using object oriented design metrics. Int. J. Innov. Res. Comput. Commun. Eng. **2**(1), 2581–2586 (2014)
9. Bandi, R., Amudhavel, J.: Object recognition using Keras with backend tensor flow. Int. J. Eng. Technol. (UAE) **7**(0), 229–233 (2018)
10. Ghuge, C.A., Ruikar, S.D., Prakash, V.C.: Query-specific distance and hybrid tracking model for video object retrieval. J. Intell. Syst. **27**(2), 195–212 (2018)
11. Rao, I.V.R., Anusha, S., Mohammad, A.B., Satish Kumar, D.: Object tracking and object behavior recognition system in high dense crowd videos for video supervision: a review. J. Adv. Res. Dyn. Control Syst. **10**(2 Special Issue), 377–380 (2018)
12. Kulkarni, J.B., Chetty, M.S.R.: Depth analysis of single view image objects based on object detection and focus measure. Int. J. Adv. Trends Comp. Sci. Eng. **8**(5), 2608–2612 (2019)
13. Krishnaveni, G., Lalitha, B.B., Vijaya Lakshmi, N.V.S.K.: An enhanced approach for object detection using wavelet based neural network. J. Phys: Conf. Ser. **1228**(1), 1742–6596 (2019)
14. Kashyap, A., et al.: Computational and clinical approach in lung cancer detection and analysis. Procedia Comput. Sci. **89**, 528–533 (2016)
15. Alami, A.E., Das, S., Madhav, B.T.P., Bennani, S.D.: Design, optimization and realization of high gain RFID array antenna 4 Ã—1 for detection system of objects in motion. J. Instrum. **14**(5), 5002–5018 (2019)
16. Harika, S., Nagarjuna, S., Naveen, T.V., Sanjay Harshanth, G., Kavya, K.C.S., Kotamraju, S.K.: Analysis of rain fade mitigation using site diversity technique in southern tropical region of India. Int. J. Eng. Technol. (UAE) **7**(1.1), 622–626 (2018)
17. Merugu, S., Tiwari, A., Sharma, S.K.: Spatial–spectral image classification with edge preserving method. J. Indian Soc. Rem. Sens. ISSN 0255-660X (2021). https://doi.org/10.1007/s12524-020-01265-7

18. Singh, O., Bommagani, G., Ravula, S.R., and Gun-Jan, V.K.: Pattern based gender classification. In: IJARCSSE, vol. 3, pp. 888–895, October 2013
19. Syed, A.T., Merugu, S., Kumar, V.: Augmented reality on sudoku puzzle using computer vision and deep learning. In: Advances in Cybernetics, Cognition, and Machine, Lecture Notes in Electrical Engineering, pp 567–578. Springer, Singapore (2020)

Chapter 12
Deep Transfer Learning for Detecting Cyber Attacks

Ragipati Karthik, Parul Shukla, S. Lavanya, L. L. Naga Satish, and J. Sai Rohith Krishna

Introduction

The web of things alludes to related gadgets sensors and actuator utilized in vehicles the electronic machines building and the structures. As the sensors records capacity and the web rise as cheap, speedier and more built in together IOT contraptions will find more noteworthy and additional reason in savvy buildings intelligent city savvy transportation frameworks and healthcare [1]. The quick advancement of IOT is to the range of life be that as it may leads to different rising cyber security dangers. This can be since IOT gadgets are frequently constrained in computing usefulness and vitality making them to be specific inclined to enemies IOT contraptions are additional uncovered to and tragically more challenging to be included from cyber assaults computers subsequently recognizing assault to protect IOT gadget from the pernicious conduct is basic to broadening the reason of IOT assault discovery method can be categorized into signature-based-techniques searching for the discover the signature of IOT ambush within the drawing closer movement these strategy requires a tall earlier know how of recognized IOT assault to characterize the signature.

The laptop learning-based methods on the same time to attempt the research the feature of normal and hostile measurements within the teaching or offline stage the predicting phase these are used to realize assault within the arriving gridlock to capabilities to mechanical and steadily study beneficial data or the knowledge aspects from amassed facts in learning procedures can become aware of various IOT result.

R. Karthik (✉) · S. Lavanya · L. L. Naga Satish · J. Sai Rohith Krishna
Department of Electronics and Computer Engineering, Koneru Lakshmaiah Education Foundation, Vaddeswaram, Andhra Pradesh, India
e-mail: karthik39r@kluniversity.in

P. Shukla
College of Law, Koneru Lakshmaiah Education Foundation, Vaddeswaram, Andhra Pradesh, India

V. Garcia Diaz and G. J. Rincón Aponte (eds.), *Confidential Computing*,
Advanced Technologies and Societal Change,
https://doi.org/10.1007/978-981-19-3045-4_12

However the computing device learning-based strategies only perform nicely below an necessary assumption the conveyance the preparing data and the predicting statistics are comparable never [2]. The less in many good application this assumption may additionally moreover now not be constantly the case in network security new kind of assaults can be positioned on a day via day basis as such the shrewd IOT records for machine getting to get it models is ordinarily exceptionally a wonderful bargain great from the data all through the preparing and offline stage to reduce the over issue an expansive volume of tutoring record with name from some IOT gadget is regularly required [3]. Be that as it may physically naming an expansive quality of truths is uncommonly time eating up and exorbitant hence deployment of computing machine learning-based procedure within the recognizing IOT attacks for an assortment of scenarios the over this work proposes a novel profound switch picking up mastery of significant trade learning strategy especially fundamentally based on auto encoder to empower in a comparative what capacities of computer picking up information in IOT assault location[4].

Related Work

Deep Transfer Learning

A deep transfer learning is the method used in the Computer technology. And that is the machine learning technique we are being used in this project [1]. By the active learning method this DTL is being used. By taking the Original Data in those data points and the features in them are collected. So by this we can get reduced data by new features.

In this DTL the raw data can be transformed in the shape of the Bytes of the data. And this data which is in bytes are the threats found in them (Fig. 12.1).

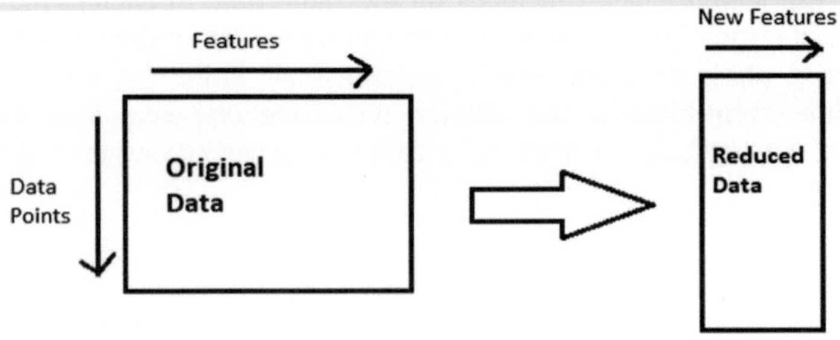

Fig. 12.1 Transfer learning

Machine Learning Techniques

Machine learning [5] is the technique in which where the humans are being depending upon the all machines in the artificial intelligence [2]. This techniques could be said that methods and approaches. And this techniques are to be said that by the computer learning this MLT is being used.

Architecture

From Fig. 12.2 the machine shape that makes use of deep transfer learning for cyber attack to detect. First, the data collection will gathers the data from all the devices which we had taken [6]. The IOT devices contains both label and unlabel information in the architecture. In the label the data is accumulated from some implement which are devoted for label data [7]. The label procedure is generally performed in the two steps for each information sample is bring out from catch seize by using Tcp trace tool then the records sample is labeled as a normal pattern or an assault test through physically analyzing the float utilizing Wireshark program [8].

The number of labeling IOT units is a better deal little than the variety of unlabeling tools. Second, the gathered facts is going to deep transfer mock up for instruction [9]. The coaching manner trying for strive to switch the expertise gathering learnt from facts with label statistics to statistics without any information [10]. The unlabeled data which we have taken can be in the form of bytes. This data can be further used to find the cyber attacks [11]. The information can be having different types of data which has no details in it. Further training, the trained deep learning mannequin is used in the detecting that cloud categorize the arriving difference from implementing for attacks in cyber as regular or assault data [12].

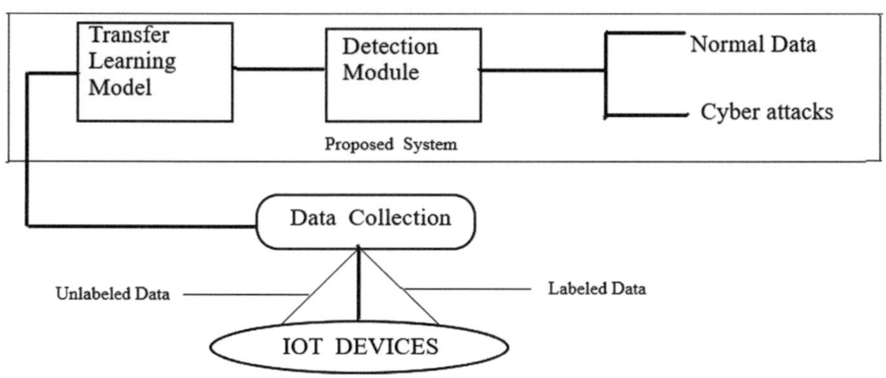

Fig. 12.2 Architecture

Cyber Attacks

Now we are knowing some cyber attacks from the following:

1. Denial of service (DOS) and disbursed denial of service (DDOS) attacks:
   The denial of service attack means that a system is unable to give the service requests. A DDOS attack is also an attack that occurs in the system program and the host machines will occur in the malicious software and to the attacker [5].
2. Man in the middle attack:
   A Mitm attack says that from server to the victim there is an attacker in the middle. The attacker get hijacks the victim computer's data. By using the sniffing technique the attacker can attack on the victim's server. The disconnection between the computer and the server is said to be the man in the middle attack.
3. Password attack:
   Many users are depending on these passwords for many reasons. But due to the hackers these are being hacked by the sniffing method. And these passwords are being hacked by the otp format also.
4. Malware attack:
   Malicious software program can be hacked by the unwanted software downloads. And by this the system may be interconnected and may hacked through this.
5. Cross-site scripting (XSS) attack:
   The XSS attack use the third party net assets to run the victims web page with the script application. The Internet site transmits the information to the attacker. So that he can get the contact with the Victim's browser.

For example, it would possibly ship the victim's server cookie, and the attacker can be extract it from the attacker's server Use it for a session hijacking. The most risky Implications arise when using XSS to take advantage of extra vulnerabilities.

In this attack the attacker discovers a website for having the script injection vulnerabilities. And the website visitor will be having that letters execute when loser dispatch his code to the assailant. The website transfer this fatality web page with the assailant freight. The fatality page can execute the spiteful letters. Attacker extracts victim's conclude. After that which he use it for the session hijacking the scripting site (Figs. 12.3 and 12.4).

6. Birthday attack:
   The birthday strike is a type of cryptology ambush that make use of the arithmetic in the back of the birthday difficulty in chance thesis. This storm can be used to the exploit imparting connecting two or extra alliance.
   This birthday ambush is an analytical circumstances applicable to the information safety that makes the brute forcing of one-way hashes easier. It is based totally off of the birthday paradox, which states that in order for there to be a

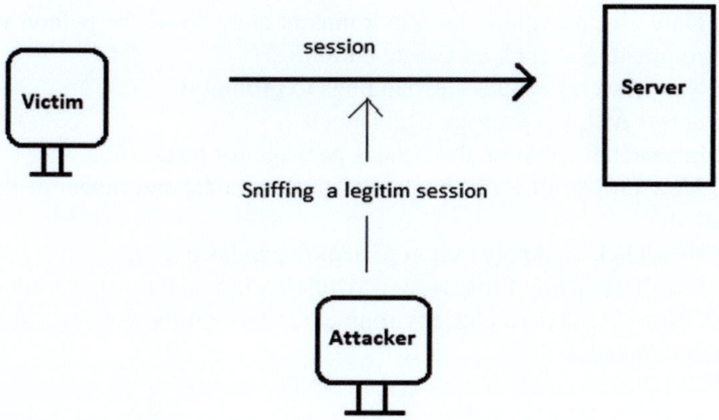

Fig. 12.3 Man in the middle attack

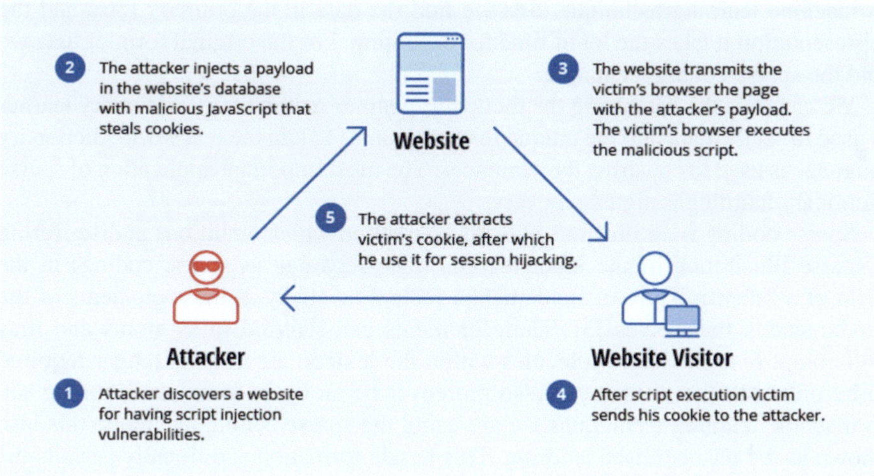

Fig. 12.4 XSS attack

50% chance that any person in a given room shares your birthday, you need 253 humans in the room.

Installing Pandas in Anaconda

Steps to Install Pandas using Anaconda Navigator:

Step1: Search for Anaconda navigator in menu bar and open it.
Step 2: Now click on the Environment tab and click on the create button.

Step 3: Now give the title to the Environment and choose the python version to the environment. and click on Create button.
Step 4: Now click on Pandas surroundings to prompt it.
Step 5: Select ALL the package deal names.
Step 6: In search bar choose the Pandas package for installation.
Step 7: Now proper click on the pandas package deal and model of the bundle installation.
Step 8: Now click on Apply button to deploy pandas package.
Step 9: Finish the set up Process by way of clicking on the Apply button.
Step 10: Now to open Pandas environment, Click on the Package to begin the pandas environment.

Sparse Coding

In machine learning technique, first we find the data in the ordinary form and the representation it takes the lot of time for executing. For the original form of data we find the sparse coding technique.

We can pass the data using the dictionary learner method. The dictionary learner is used to set the data into the unique form of atoms [13]. In the real world, dictionary language is used to construct the sentences. The most important application of sparse dictionary learning is signal recovery.

Sparse coding is an illustration learning method which ambitions at discovering a sparse illustration of the input records (also regarded as sparse coding) in the form of a linear mixture of fundamental factors as nicely as these elements of the fundamentals themselves [3]. These elements can referred to as atoms and they will compose a glossary. Molecules within the lexicon are now no longer required to be orthogonal, and they can also moreover be an over- complete spanning set. In machine learning techniques we are using the sparse coding technique this also known as the unsupervised learning. This hassle formation additionally permits the proportions of the indicators being taken to be highest than the only one of the signals being observed [14]. The following are two houses contributed to having the seemingly redundant atoms that permit more than one representations of the identical signal however additionally furnish an enhancement in sparsity and flexibility of the representation.

In compressed sensing, a high-dimensional sign are often recovered with only a couple of linear measuring the long as the sign is sparse or nearly sparse [4]. Since now not all signals satisfy this sparsity condition, it is of brilliant significance to discover a sparse representation of that signal such as the wavelet transform or the directional gradient of a raster zed matrix.

In one of the ideas of dictionary studying is that the dictionary has to be inferred from the enter data. The emergence of sparse dictionary getting to know techniques was once motivated by way of the truth that in sign processing one normally wants to characterize the enter facts the use of as few factors as possible [15]. However, in certain cases a dictionary that is trained to suit the input information can drastically improve the sparsity, which has applications in information decomposition, compression and evaluation and has been used in the fields of photo denoising and classification, video and audio processing. Sparsity and over complete dictionaries have huge applications in picture compression, picture fusion and in painting [16].

Parametric training strategies are aimed to comprise the fine of each worlds—the realm of analytically built dictionaries and the realized ones. This allows to the more powerful generalized workbook that can be utilized to the instances of arbitrary-sized sign. Notable strategies include:

Translation-invariant dictionaries: These dictionaries are composed with the help of the translations of the particles from the lexicon that were constructed for a finite-size sign patch. The dictionaries can be transferred to the elements in the sparse coding techniques. Mainly this may lead to the other strategies in which the translations of the training strategies.

Multi scale dictionaries: For improving the sparsity in dictionaries which are focusing on the construction of the scaled dictionaries.

Sparse dictionaries: The technique is focusing on not only a sparse representation but additionally developing the sparse dictionary which is developed by the expression of $D = BA$ Where B is some pre-defined analytical dictionary with applicable houses such as speedy computation and A is a sparse matrix. These components allows us for without delay combine the quickly presenting the analytical dictionaries with the flexibility of sparse approaches in it [17].

Dictionary Learning

Dictionary studying is a department of signal processing and computer mastering that targets at finding a body is called dictionary in which some of the training facts admits a sparse representation. The sparser the representation is the higher than the dictionary.

Efficient dictionaries: In the ensuing dictionary is in common a dense matrix, and its manipulation can be computationally and costly both at the stage and later in the use of this dictionary, for duties such as sparse code techniques [11]. Dictionary getting to know is therefore limited to fantastically small-scale problems. Inspired by means of regular fast transforms, we proposed an ordinary dictionary structure that permits cheaper manipulation, and an algorithm to study such dictionaries—and their speedy implementation—over coaching data.

Sample complexity: Besides dictionary learning, many present day tools in computer getting to know and sign preparing depend on the factor of a matrix which is got by concentrating on the dimension of the vectors from a coaching collect. In which the factor undertaking would be to reduce the anticipated first-rate of the factors which are under the distribution of the coaching vectors, it is to realize or execute in exercise via minimizing an observed common over the viewed collection [3]. In this Complexity of Dictionary space Learning and different Matrix outcomes, we can provide sample complex estimates to uniform and manage how a good deal the empirical standard diversions from the predicted price function.

Provably precise dictionary learning: When besides pattern complexity issues, and vital theoretical problem is to represent the minimal of the non-convex price functions designed for dictionary learning. In Sparse and dictionary studying with noise and out layers we establish that, with high possibility, sparse coding admits a close-by the negligible spherical in the reference of dictionary generating the signals [4].

The evaluation is non-asymptotic and highlights the function of the key portions of the solved, such as the consistency, the degree of the signal diversions, the number of molecules or atoms, the sparsity, and the range of perceptions.

Simply put dictionary gaining knowledge of is the technique of learning a matrix, known as a dictionary, such that we can write a sign as a linear aggregate of as few columns from the matrix as possible.

When using dictionary getting to know for images we take benefit of the property that natural photographs can be represented in a sparse way. The potential that if we have a set of simple picture elements for any picture. It can be composed as a straight aggregate of solely a few simple features. The matrix we name a dictionary is such a set. Each column of the lexicon is one fundamental photo features. These characteristic vectors are known as atoms in literature [12].

Comma Separated Values

The file is in the form of a Comma Separated Values is an easy textual content file or the record that accommodates a listing of facts. These archives are commonly used the altering statistics between specific advantages. Let us take the example of databases and school files regularly useful resource CSV files.

These files can also from time to time be acknowledged as the Comma Delimited files. There in the important use the comma persona to separate the data, however now and again we use exclusive characters, like semicolons or the alphabets. The notion is that you can transfer the problematic information from the utility to a csv classify and the particulars is imported in that folder into each and every other applications in it.

The structure of a CSV file:

The structure of aport folio can surely be greater difficult than anyone can include hundreds of strokes, extra appearance on every rule, or long strings of the text. The CSV archives has additionally now no longer even had the headers on the up and some may also have additionally use citation to encompass every bit of facts, then again that is the fundamental layers [6].

That simple feature can use the CSV archives are outlet to give a way to without difficulty transfer information and detract it into different outputs. This output details is mortal clear and can be effortlessly considered with a textual content deskman like the notepad or array (matrix) utility like Excel.

How to import CSV file:

However, many CSV archives are made for importing into one of a kind of programs. You too send out your contacts from saved Contacts on the Google, your passwords from Last Pass, or a large quantity of documents from a database or the dataset program. Then the output CSV archives can be a signification into functions in guide form of the statistics [11].

Depend on the utility from in which you are transferring the data, you may got to choose out an first-rate CSV structure for the aim implementation. Example, saved contacts can transfer the contacts in both Google and Outlook for Outlook layout. Whichever way you get a folder and can accommodate the facts, then again it is prepared in a barely one of a kind way. In a gorgeous application, seem for the "Import" or "Import CSV" alternative which can permits you to pick the file to detract in the view point we will choose the file, then click on the file, open the file, then import/ export, then the contacts are transferred from the csv file.

How to open a CSV file in excel:

If you installed the excel you simply digitize on a folder to extend it in the surpass. After digitizing the folder you possibly look at an instantaneous asking the software yourself to choose and select the excel.

If you opened the excel then you can take the csv file. If you do not see the folder then you exist in favor of open, you might also change the file named in sort to open the "Text Files". Then shine will show the information in a notepad [17].

Then you barrel transfer the information out from a folder upon current spreadsheet.

In information lappet within the get and change information batch, click from text or csv.

In drift data frame, paired the file you choose to purport or transfer and click on detract.

In premier trunk you will have many possibilities like:

First stack which you prefer to the information without delay to the spreadsheet.

Second choose stack to if they desire to bundle the statistics to a desk or the grind page.

Third change the facts you desire to pack the information to ability model and improve it earlier than starting it in the excel.

Results and Outputs

In Anaconda we have executed the code. In that code first in the excel sheet we have taken the input data. In the input data we have taken the data as the server number, bytes of data and number of clients in it. So after running the program we found the threats in it in the form of the bytes of data (Figs. 12.5 and 12.6).

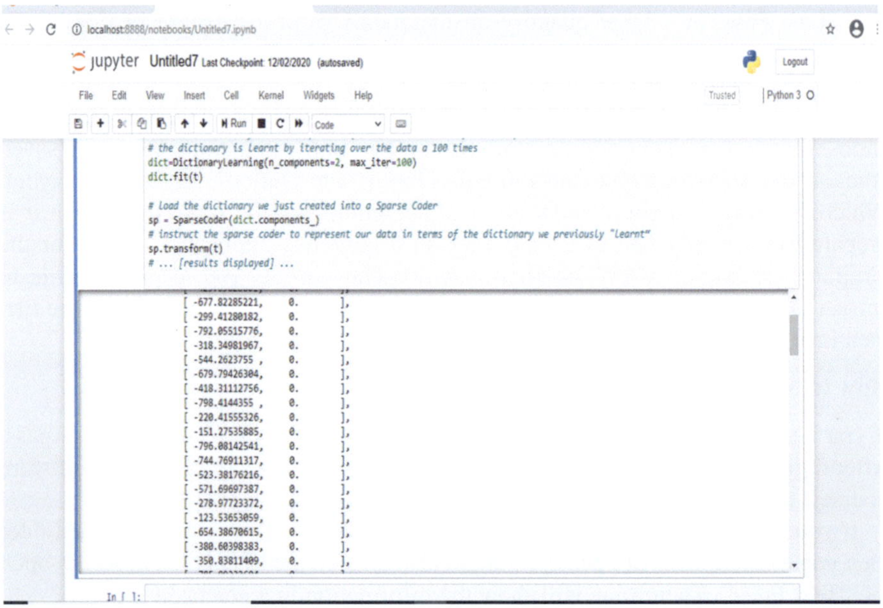

Fig. 12.5 Output shown in bytes

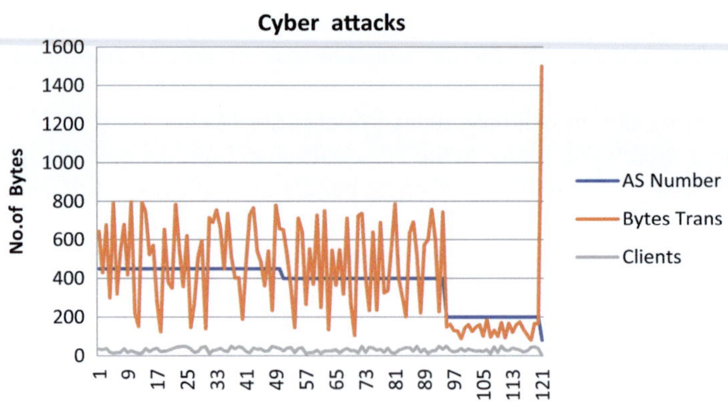

Fig. 12.6 Graph of attacks in bytes

Conclusion

In this project by using the machine learning process technique we noticed that the data which is taken are in bytes. By using the CSV file within the comes about the attacked data is in bytes. Finally by using this AI procedures we have taken note the attacks and this can be helpful to the others in future that how much of data has been attacked.

By this Project we are just finding the attacks which are noticed from the server and from the client data. And in future we can find and name the each and every attack from this project.

References

1. Luong, N.C., Hoang, D.T., Wang, P., Niyato, D., Kim, D.I., Han, Z.: Data collection and wireless communication in Internet of Things (IoT) using economic analysis and pricing models: A survey. IEEE Commun. Surveys Tuts. **18**(4), 2546–2590 (2016)
2. Meidan, Y., Bohadana, M., Shabtai, A., Ochoa, M., Tippenhauer, N.O., Davis Guarnizo, J., Elovici, Y.: 'Detection of unauthorized IoT devices using machine learning techniques (2017) arXiv:1709.04647. [Online]. Available: http://arxiv.org/abs/1709.04647
3. Nisioti, A., Mylonas, A., Yoo, P.D., Katos, V.: From intrusion detection to attacker attribution: A comprehensive survey of unsupervised methods. IEEE Commun. Surveys Tuts. **20**(4), 3369–3388 (2018)
4. Jyothi, B., Sruthi, D., Karthik, R.: Energy monitoring using arm 7. Int. J. Innov. Technol. Exploring Eng. (IJITEE) **8**(6), 441–445 (2019)
5. Ahmed, M., Ahmed, R., Thakuria, A.J., Laskar, R.H.: Eye center guided constrained local model for landmark localization in facial image. In: 2019 9th Annual Industrial Automation and Electromechanical Engineering Conference (IEMECON). IEEE (2019)
6. Meidan, Y., Bohadana, M., Mathov, Y., Mirsky, Y., Shabtai, A., Breitenbacher, D., Elovici, Y.: N-BaIoT—Network-based detection of iot botnet attacks using deep autoencoders. IEEE Pervasive Comput. **17**(3), 12–22 (2018)
7. Vlajic, N., Zhou, D.: IoT as a land of opportunity for DDoS hackers. Computer **51**(7), 26–34 (2018)
8. Shaik, A.S., Bhavani, M., Ravi Kiran, K.: Smart pick and drop intimation system of school children. Ind. J. Sci. Technol. **10**(46), pp. 1–7 (2017). https://doi.org/10.17485/ijst/2017/v10i46/117158. ISSN(Print): 0974-6846, ISSN(Online): 0974-5645
9. Gow, R., Rabhi, F.A., Venugopal, S.: Anomaly detection in complex real world application systems. IEEE Trans. Netw. Service Manage. **15**(1), 83–96 (2018)
10. Khattak, S., Ramay, N.R., Khan, K.R., Syed, A.A., Khayam, S.A.: A taxonomy of botnet behavior, detection, and defense. IEEE Commun. Surveys Tuts. **16**(2), 898–924 (2014)
11. Bahsi, H., Nomm, S., La Torre, F.B.: Dimensionality reduction for machine learning based IoT botnet detection. In: Proceedings of 15th International Conference Control, Automation, Robotics and Vision (ICARCV), pp. 1857–1862 (2018)
12. Dromard, J., Roudiere, G., Owezarski, P.: Online and scalable unsupervised network anomaly detection method. IEEE Trans. Netw. Service Manage. **14**(1), 34–47 (2017)
13. Buczak, A.L., Guven, E.: A survey of data mining and machine learning methods for cyber security intrusion detection. IEEE Commun. Surveys Tuts. **18**(2), 1153–1176 (2016)
14. Trinath Basu, M., Karthik, R., Mahitha, J., Lokesh Reddy, V.: IoT based forest fire detection system. Int. J. Eng. Technol. **7**(2.7), 124–126 (2018)

15. Krishnaveni, K., Venkata Ratnam, K., Prathyusha, G., Gopi Krishna, P.: Development of real time environment monitoring system using with MSP430. Int. J. Eng. Technol. (UAE) **7**(2), 72–76 (2018)
16. Kandaswamy, C., Silva, L.M., Alexandre, L.A., Sousa, R., Santos, J.M., de Sa, J.M.: 'Improving transfer learning accuracy by reusing stacked denoising autoencoders. In: Proceedings of IEEE International Conference System, Man and Cybernetics (SMC), pp. 1380–1387 (2014)
17. Zhuang, F., Cheng, X., Luo, P., Pan, S.J., He, Q.: Supervised representation learning: transfer learning with deep autoencoders. In: Proceedings of 24th International Joint Conference on Artificial Intelligence, pp. 4119–4125 (2015)

Chapter 13
Data Security in Cloud with Hybrid Homomorphic Encryption Technique Using GM–RSA Algorithm

Pachipala Yellamma, Uppalapati Sai Santosh, Repala Yashwanth, Tatikonda Uday Amartya Sai, and Garimella Naga Sai Uma Sampath Kumar

Introduction

Presently, days are the times of cloud computing in the IT industries. Like authentic fogs which are the social events of water particles, the term "cloud" in distributed computing is that the assortment of organizations? The cloud figuring is a utilization Internet as a premise having loads of organizations as the most impressive design of calculation [1]. The client can now ready to use the modalities of conveyed processing immensely at whatever point mentioned. Rather than fixing their own real structure, the customers customarily favour a centre individual provider for the organization of the Web. The essential examples of distributed computing which are utilized by broad individuals in way of life are Facebook, YouTube, Dropbox, Gmail and so forth. It offers adaptability, adaptability, dexterity and effortlessness that are the reason its utilization is quickly expanding in the ventures.

"Not all kinds of cloud computing raise an equivalent privacy and confidentiality risks. Some accept that a significant part of the computing action happening today totally on PCs possessed and controlled locally by clients, which will move to the cloud within the future, by connecting to several servers located on organization premises [2]. Public and private partnerships are lately a normally received example of governance to satisfy the various needs of their citizens confidently and providing quality of those services. Cloud computing technology acts as a mediator between public and personal partnerships. In such cases, there is a good opportunity that an outside party are often involved in providing cloud services having control over the

P. Yellamma (✉) · U. S. Santosh · R. Yashwanth · T. U. Amartya Sai · G. N. S. U. Sampath Kumar
Department of CSE, Koneru Lakshmaiah Education Foundation, Vaddeswaram, Andhra Pradesh, India
e-mail: pachipala.yamuna@gmail.com

V. Garcia Diaz and G. J. Rincón Aponte (eds.), *Confidential Computing*, Advanced Technologies and Societal Change,
https://doi.org/10.1007/978-981-19-3045-4_13

information on processing and transmission of data and security rules. With innovation of new instruments and advances, the assailants are planning new strategies to avoid present security models and cause a security break [3]. There are many cloud servers' farms which store an extremely enormous measure of data from sources and backing information driven calculation. Protections are regularly significant issues for such server farms when the information they need is delicate. A server farm could likewise be assaulted, traded off and add the capability of insider assaults [4]. The eventual outcome of which is at this point encoded information, notwithstanding, it can again unscramble back to the results of the (reasonably) same capacities applied to the plain information.

Homomorphic Encryption

Because of the protection spillage of touchy information, the traditional encryption framework is not totally secure from a mediator administration like cloud workers [5]. HE is a special sort of encryption component that can resolve the issues of well-being and security from the cloud. In contrast to the public key encryption, which has three security systems, i.e. key age, encryption and decoding; there are four methods in HE plots, including the assessment calculation as demonstrated in Fig. 13.1. The HE permits the specialist organization (outsider) to play out specific kinds of procedure on the client's encoded information without decoding the scrambled information while keeping up the security of the clients' encoded information. In homomorphic encryption, the client needs to store information in the cloud, the client can encode the data and arrangements the scrambled data onto the cloud. The CS runs a calculation on scrambled information utilizing HE elite of huge substance of the encoded information. At that point, CS goes before the scrambled information back, and the client is furnished with a key to unscramble the encoded information while saving the insurance of his data as demonstrated in Fig. 13.1. In HE, numerical procedure on the plaintext during encryption is comparable to another activity performed on the ciphertext. Allow us to consider a basic homomorphic procedure on plaintext amongst an identical code text activity.

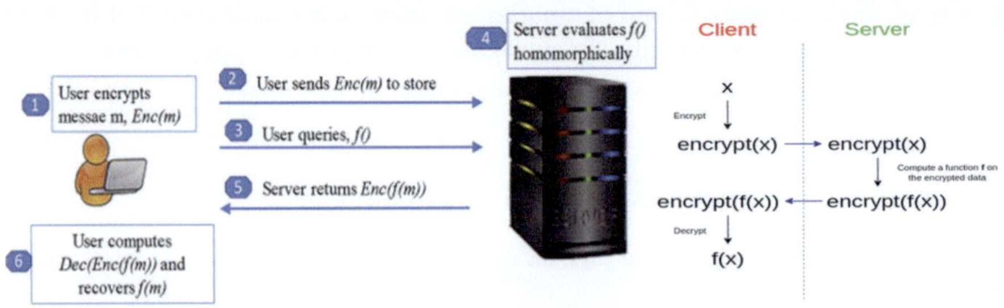

Fig. 13.1 Homomorphic encryption (HE) scenario

Let $E(m_1) = m_1^e$ and $E(m_2) = m_2^e$,
Then,
Addition Homomorphism:
$E(m_1) + E(m_2) = m_1^e + m_2^e = (m_1 + m_2)^e = E(m_1 + m_2)$
Multiplication Homomorphism:
$E(m_1) \times E(m_2) = m_1^e \times m_2^e = (m_1 \times m_2)^e = E(m_1 \times m_2)$

The overall technique of the HE is:

Creation of key: Secret key *Ks* and public key *Kp* $(Ks, Kp) = \text{keyGen}(S)$.

Encryption: By using encryption technique, K_p is obtained and PlainText(M) is scrambled and provided the cipherText(C) $C = \text{Encpk}(M)$.

Evaluation: Apply a capacity f towards cipherText(C) utilizing the public key $C^* = \text{Evalpk}(f, c)$.

Decoding: Calculation gets the cipherText(C) and the private-key recuperates PlainText(M) $M = \text{Decsk}(C)$.

Types of HE

1. Partially-Homomorphic-Encryption: In this plan, the numerical activity is permitted on the encoded message, i.e. either expansion or duplication activity, with a limitless number of times.
2. Somewhat-Homomorphic-Encryption: In this plan, both expansion and duplication activities are permitted, however, with just for the predetermined number of times.
3. Fully-Homomorphic-Encryption: In this plan, it permits various kinds of assessment procedure on the scrambled message with a limitless number of times.

Homomorphic Encryption Techniques

Many different types of techniques that are available in homomorphic encryption, in this paper, we are presenting the two promising algorithms; the greatly effective in this era of cloud computing for securing our privacy on cloud without being exposing the data even to the service providers is as follows:

A. GM—Additive PHE: GM algorithm is as shown. Expect we should need to encode two pieces: $m1$ and $m2$ utilizing the GM cryptosystem.

$$\begin{aligned}\text{Additive} &= Enc(GM(m1)) \times (GM(m2)) \\ &\equiv ((bm1 \times r12) \cdot (bm2 \times r22)) \bmod n \\ &\equiv bm1 + m2(r1r2)2 \bmod n \\ &\equiv EncGM(m1 \oplus m2, pk)\end{aligned}$$

B. RSA—Multiplicative PHE.

RSA, in 1978, published their public key cryptosystem. Despite the fact that it is a fundamental algorithm, but it is considered the critical piece of homomorphic encryption, because of this it acts as multiplicative encryption technique. Functioning of RSA algorithm is as shown below. Homomorphic encryption property of the RSA is assume two CipherTexts($C1$, $C2$)

$$C1 = m\,1\mathrm{e} \bmod n \quad C2 = m\,2\mathrm{e} \bmod n \quad C1 \cdot C2 = (m\,1\mathrm{e} \cdot \mathrm{m2e}) \bmod n$$

It is evident that RSA is an essential algorithm with the restricted computation choice. It is really reduces the uses of the algorithm. But a significant algorithm since it goes about while a structure block and numerous improved algorithms depend on Rivest, Shamir and Adleman algorithm. It is also important to obtain that RSA is speedy and it can practically implement.

Literature Survey

Gentry et al. [6] developed a basic public key encryption plot which upholds the polynomial, i.e. numerous increases and one augmentation, like the cryptosystem of Boneh, Goh and Nissim (BGN), referred to be just about as hard as certain most pessimistic scenario grid issues. A few highlights of the cryptosystem incorporate help for enormous message space, a simple method of accomplishing recipe protection, a superior message-to-encode text extension proportion than BGN and a simple method of increasing two scrambled polynomials. Likewise, the plan can be made character based and spillage strong. They depicted an additively homomorphic encryption conspire, and furthermore upholds one increase. This issue, which is identified with the notable "learning equality with commotion", has gotten standard in the investigation of grid-based cryptography.

Sharma et al. [7] proposed privacy homeomorphisms which provide a way for the security of information that should be worked on. They are of characteristically restricted relevance, since examinations may not, as a rule, be remembered for the arrangement of activities to be utilized. One of the essentials is clearly inborn, and restrictions of the method are that a data framework chipping away at the encoded structure an information can generally store or improve the information for the client, any more convoluted activities require the information to be decoded prior to being worked on. In any case, there are a couple of genuinely inborn limits on what can be cultivated [8]. These extraordinary encryption capacities we call "protection homeomorphisms", they structure a fascinating subset of subjective encryption plans (called "security changes").

Paillier [9] The Paillier encryption plot is a probabilistic uneven calculation for public key cryptography conspire calculation utilizing decisional composite degree residuosity issue. It is an added substance homomorphic cryptosystem plot; it implies that lone the public key and the encryption message ($m1$ and $m2$), that be able to be registered as the encryption message ($m1 + m2$). It is embraced by the safe scalar item and it has been generally utilized in protection saving information mining. Another element named self-blinding—it is the capacity to change a ciphertext into another without modifying any of the substance it is decoding [10].

Yellamma et al. [11] introduced a homomorphic public key encryption depends on limited gatherings of composite request which upholds a bilinear guide. Utilizing a development as per Paillier, they get a framework with an added substance homomorphism [12, 13]. As the yield, our frameworks support discretionary increments and duplication on scrambled information. This property, thus, permits the assessment of multiplicative polynomials of all out degree 2 on scrambled qualities [14]. The protection of this plan depends on another hardness supposition and subgroup choice issue. To be specific, the given component of the composite gathering request $n = q1q2$, this is unusable to choose whether it has a place with a subgroup of request $q1$. As an immediate use of the new homomorphic encryption conspires, a convention is developed for negligently assessing 2-DNFs [15]. The convention gives a quadratic improvement in correspondence intricacy over confused circuits [16]. The convention told the best way to get a private data recovery conspire (PIR) as a variation of the 2-DNF convention. As verified over, the encryption plot is utilized to assess quadratic multivariate polynomials on ciphertexts gave the resulting appraisal course inside a little set; specifically, to process speck items on ciphertexts. What's more, this property is utilized to make a contraption that empowers the check that encrypted% esteem is one of two "great" values. The device is% used to develop an effective political race convention where electors do not have to give evidence of vote legitimacy. Finally, improvement of the political decision technique to a system of generally irrefutable calculation is finished.

Proposed Methodology

Work measure is the plan of activities that are imperative to complete a task. Every movement in a work cycle has a specific development before it and a specific development after it, beside the underlying advance. In a straight work measure, the underlying advance is by and large begun by an outer event. Interaction guides and flowcharts are successful instruments for imagining the number and request of steps in a work process. Flowcharts utilize straightforward mathematical images and bolts to characterize in the event that/connections. Cycle maps look fairly comparative, yet they incorporate data, archiving the assets that each progression in an interaction requires. Here, the proposed workflow will be as follows:

The workflow of proposed method.

Phase I: Load unique information (plaintext).

Phase II: Convert the information to Binary.

Phase III: Select arbitrary massive number prime's number p and q.

Phase IV: Encryption using GM algorithm and generates ciphertext C1.

Phase V: RSA cryptosystem key creation.

Phase VI: Encryption using RSA algorithm and generates ciphertext C2.

Phase VII: Upload the scrambled information to the cloud and interaction (Add-multiplication).

Phase VIII: Decryption of the ciphertext C2 produces ciphertext C1.

Phase IX: Decryption of the ciphertext C1 produces unique prepared information.

1. Encryption using GM algorithm
2. Encryption using RSA algorithm
3. Performing operations on data
4. Decryption using RSA algorithm
5. Decryption using GM algorithm

Encryption using GM algorithm: The plaintext which is has to be stored or computed in cloud is changed into binary format. These binary digits are now encrypted using GM algorithm which is additively homomorphic, but it encrypts just only a single bit. Thus, it provides data confidentiality. This module helps in key generation and encryption of data using GM algorithm.

Encryption using RSA algorithm: The plaintext which is encrypted using GM algorithm and again RSA algorithm. RSA is considered as effective piece of homomorphic encryption and because of this reason it has been incorporated to act as an example of MHE technique (multiplicative homomorphic encryption). This module does key generation and encryption of data using RSA algorithm.

Performing operations on data

Now the scrambled information is put away in cloud, and tasks are performed on encoded information without decoding it.

$$\begin{aligned} E(m1) * E(m2) &= (y_1^2 b^{m1} (\text{mod } N)) * (y_2^2 b^{m2} (\text{mod } N)) \\ &= (y_1 * y_2)^{2b(m1+m2)} (\text{mod } N) \quad = E(m1,\, m2) \end{aligned}$$

RSA algorithm with multiplicative property is given as:

$$C1 \equiv m1^E (\text{mod } N) \quad C2 \equiv m2^E (\text{mod } N)$$

By multiplying ciphertext of two messages $C1 \cdot C2 \equiv (m1^E \cdot m2^E) \bmod n \equiv (m1 \cdot m2)^E (\text{mod } N) = E(m1 \cdot m2)$.

Gives the same ciphertext or product of messages, i.e. $m1$ and $m2$.

The result produced is again in encrypted form which when decrypted gives same result as performed on raw data.

Decryption

The ciphertext produced is decrypted again using GM and RSA algorithms to produce the result in plaintext.

GM Algorithm

1. ***Key Generation:***

   I: The client is produced with public key or private key pairs by selection of two huge irregular prime numbers P and Q.

   II: Calculate $N = P \cdot Q$ select b has place with $B_n|b$ (is a quadratic value modulo n and $(b/n = 1)$).

   III: The key pair is taken as (pk, sk) where

$$pk = (n, b)\{\text{publickey}\} \quad \text{and} \quad sk = (p, q)\{\text{privatekey}\}$$

2. ***Encryption***: Information message M comprises various pieces $m1, m2, \ldots mn$

   Stage I: For each piece mi, an arbitrary worth y_i is produced | $\gcd(y_i, N) = 1$.

   Stage II: Compute $C_i = y_i 2bm_i (\text{mod } N)$.

3. ***Decryption***:

   Scrambled message C comprises numerous pieces $(C1...Cn)$. Recuperate m utilizing the accompanying technique. For every i, utilizing the excellent factorization (p, q) decides if the worth C_i is a quadratic build-up provided that this is true, $m_i = 0$, in any case, $m_i = 1$. The message $m = (m1,\ldots, mn)$.

RSA Algorithm

1. **Key Generation**:

I: The customer makes a private or public key dependent on picking two irregular enormous prime number P and Q. Euler's totient $\phi(N) = (P - 1)(Q - 1)$.

II: Choose the encryption key E randomly Where, $1 < E < \phi(N)| \gcd(E, \phi(N))$ and decryption key D as $D \equiv E^{-1}(\text{mod } \phi\, (N))$;

III: Calculate public key (Kp) and private key (K_s):

$$\text{publikkey}(kp) = \{E, N\} \quad \text{and} \quad \text{privetekey}(ks) = \{D, P, Q\}$$

Experimental Results and Analysis

Methods Compared: Analysis and simulations are done on the hybrid algorithm based on comparison of encryption time of dissimilar plaintext extent and comparison of decryption time of dissimilar ciphertext extent.

Metrics Calculated.

1. The encryption-time in seconds of enhanced homomorphic encryption (EHE) algorithm.
2. The encryption-time seconds of proposed algorithm.
3. Decryption-time in seconds of enhanced homomorphic encryption algorithm.
4. Decryption time in seconds of proposed algorithm.

I. **Comparison of encryption time of dissimilar plaintext extent**

The examination is finished by picking three content records of various sizes, which inputs are given for the calculations to affirm the exhibition of the GM–RSA half breed homomorphic encryption cryptosystems. Table 13.1 and Table 13.2 show a various plain text size for the EHE calculation which support one activity multiplicative interaction and the presented calculation which upholds two tasks added substance multiplicative cycle multiple times quicker. Table 13.1 and Fig. 13.2 show the plaintext size, encryption time in second for the proposed calculation contrasted and EHE.

From Table 13.1, charts are drawn to signifies the time taken by the acquainted calculation and EHE with encode various sizes of scrambled information.

II. **Comparison of decryption time of different cipher text sizes**

On a comparative machine, diverse ciphertext sizes for the EHE algorithm support one activity multiplicative cycle, and the presented algorithm upholds two tasks

Table 13.1 The plaintext size and encryption time for proposed calculation contrasted and EHE

Plaintext size in bits	Encryption time in second of EHE	Encryption time in second of proposed algorithm
8	0.0724	0.0245
16	0.1093	0.0693
24	0.1308	0.0815
32	0.1429	0.0885
48	0.1589	0.0915
64	0.1696	0.0985

Table 13.2 Ciphertext and decryption time for proposed algorithm compared with EHES

Ciphertext size in bits	Decryption time in second of EHE	Decryption time in second of proposed algorithm
8	1.86	0.93
16	5.07	2.12
24	6.57	3.35
32	6.87	3.89
48	8.94	4.86
64	10.72	5.97

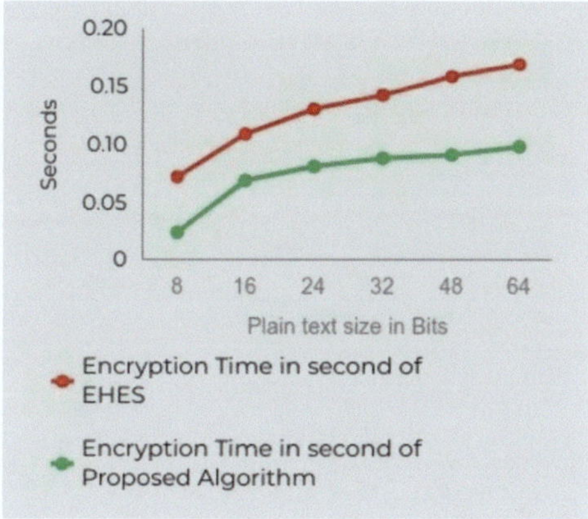

Fig. 13.2 Plaintext size versus encryption time in second for proposed algorithm compare with EHE algorithm

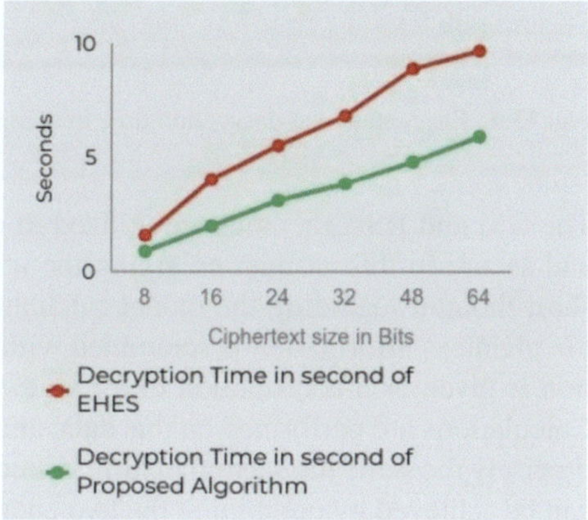

Fig. 13.3 Plaintext size versus decryption time in second for proposed algorithm compare with EHE algorithm

added substance multiplicative interaction and their decoding times are looked at. The comparison is given in Table 13.2. The decryption time is also less for hybrid algorithms when compared to the EHE algorithm.

From Table 13.2, graphs are drawn which represent the time taken with the projected algorithm and EHE to decrypt different sizes of encrypted data (Fig. 13.3).

In Fig. 13.4, encryption and decryption time in second of hybrid EHE proposed algorithm is better performance than compared to the EHE

Conclusion

A new technique of combining two PHE systems to form a single fully homomorphic encryption is developed. The combination of advantages that are observed in the two standalone techniques, a half breed homomorphic encryption calculation is created.

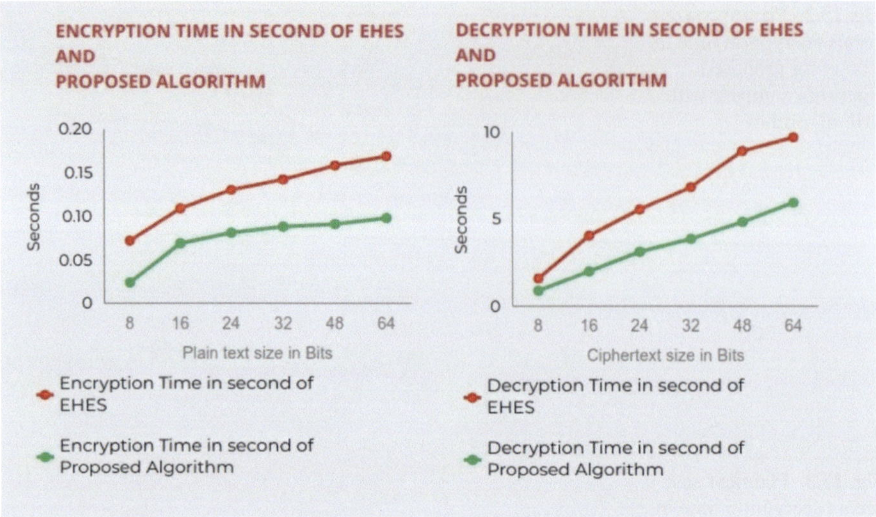

Fig. 13.4 Encryption and decryption time in second of hybrid EHE proposed algorithm

The GM and RSA methods are utilized to give upgraded information classification and secure in this manner enlarging the utilizations of such procedures. To give a short thought regarding the model calculation utilized in a technique to encourage the plaintext information is scrambled with GM at the start. This encoded information is given as a contribution to a framework that further scrambles it with RSA. Calculations are performed on this data, and the technique for unscrambling follows precisely the same interaction in turn around request. Reduction in computation time can be achieved by combining the two encryption algorithms.

References

1. Yellamma, P., Narasimham, C.: Hybrid compressed hash based homomorphic AB encryption algorithm for security of data in the cloud environment. Int. J. Future Revolution Comput. Sci. Commun. Eng. **3**(11), 604–612 (2017)
2. El Gamal, T.: A public key cryptosystem and a signature scheme based on discrete algorithms. IEEE Trans. Inform. Theory **31**(4), 469–472 (1985)
3. Gentry, C.: Fully homomorphic encryption using ideal lattices. Proc. Forty-First Ann. ACM Symp. Theory Comput. **1**, 169–178 (2009)
4. Gentry, C.: Toward basing fully homomorphic encryption on worst case hardness. Adv. Cryptol.—CRYPTO 2010 **62**, 116–137 (2010)
5. Yellamma, P., Dileep Kumar, P., Sai Pradeep Reddy, K., Sri Harsha, L., Jagadeesh, N.: Probability of data leakage in cloud computing. Int. J. Adv. Sci. Technol. **29**(6), 3444–3450 (2020)
6. Gentry, C., Halevi, S., Vaikuntanathan, V.: A simple BGN-type cryptosystem from LWE. In: Gilbert, H. (ed.) Advances in Cryptology—EUROCRYPT 2010. Lecture Notes in Computer Science, vol. **6110**, pp. 506–522. Springer, Berlin (2010)

7. Sharma, S.K., Jain, K., Suresh, M.: Quantitative evaluation of panorama softwares. Lecture Notes in Electrical Engineering, vol. 500, 543–561 (2019)
8. Rivest, R.L., Adleman, L., Dertouzos, M.L.: On data banks and privacy homomorphisms', foundations of secure computation. Acad. Press **1**, 169–179 (1978)
9. Paillier, P.: Public-key cryptosystems based on composite degree residuosity classes. EUROCRYPT '99 **1**, 223–238 (1999)
10. Boneh, D., Gah, E.-J., Nissim, K.: Evaluating 2-DNF formulas on ciphertexts. Theory Crypt. Conf. TCC'200S **3378**, 325–341 (2005)
11. Yellamma, P., Rajesh, P.S.S., Pradeep, V.V.S.M., Manishankar, Y.B.: Privacy preserving biometric authentication and identification in cloud computing. Int. J. Adv. Sci. Technol. **29**(6), 3087–3096 (2020)
12. Bhavani, M., Narayana, V.A., Sreevani, G.: A novel approach for detecting near-duplicate web documents by considering images, text, size of the document and domain. Lecture Notes in Electrical Engineering, vol. 398, pp. 1355–1366 (2021)
13. Tebba, M., El Hajji, S., El Ghazi, A.: Homomorphic encryption applied to the cloud computing security. In: The Proceeding of World Congress on Engineering, vol. 1, pp. 614–620. London, UK (2012)
14. Sonth, M.V., Srikanth, G., Agrawal, P., Premalatha, B.: Basic logic gates in two dimensional photonic crystals for all optical device design. Int. J. Electron. Telecommun. **67**(2), 247–261 (2021)
15. Debnath, S., Islam, M.: Disinfection chain: a novel method for cheap reusable and chemical free disinfection of public places from SARS-CoV-2. ISA Transactions (2021)
16. Shaik, A.S., Usha, S.: Sensor based garbage disposal system. Int. J. Innov. Technol. Exploring Eng. **8**(452), 164–167 (2019)

Chapter 14
Pragmatic Reform to Ameliorate Insider Data Theft Detection

K. Sri Sumanth, Sridevi Sakhamuri, K. Yaswanth, K. Revanth, and L. Chandra Sekhar Reddy

Introduction

Insider data thefts are increasing significantly and they are hazardous to any organization. The IT and security teams like blue team are hunting for the reasons behind those attacks. One of the main cause of the attack are the reckless and irresponsible employees. They either raise the attack vector or they might be behind the attack by supporting the hackers. The main problem is that many employees in the organization has proper and legitimate access to the assets and data. This makes the insider theft detection and prevention more challenging than external threats. Plenty of people are doing enormous research about insider threats and theft.

Insider Threat Definition

An insider threat can be defined as evil threat that is caused by an individual who is an employee or a former employee or then who has a legitimate access to the assets of an organization and they result bad outcomes to an institution, a company or an organization.

K. Sri Sumanth · S. Sakhamuri (✉) · K. Yaswanth · K. Revanth
Department of ECM, Koneru Lakshmaiah Education Foundation, Guntur, Andhra Pradesh, India
e-mail: sridevisakhamuri@kluniversity.in

L. Chandra Sekhar Reddy
Department of CSE, CMR College of Engineering and Technology, Hyderabad, Telangana, India

V. Garcia Diaz and G. J. Rincón Aponte (eds.), *Confidential Computing*,
Advanced Technologies and Societal Change,
https://doi.org/10.1007/978-981-19-3045-4_14

Insider Data Theft Definition

An insider data theft can be defined as stealing of data, assets or any sort of valuable information of a company or an organization that is stored in a server, data based or in other storage devices. It may be done directly by an insider or a compromised employee is used as an intermediator to perform an attack.

Insider Threat Versus Insider Data Theft

The insider threats are those caused by an employee which makes the organization vulnerable. Among them, one of the possible menace is data theft. This causes huge loss of money, trust, reputation, compromise of customer's data, disclosure of trade secrets and also to face legal liabilities. The compromised employees accounts lead the cybercriminals to lurk inside a network and stay unnoticed for longer periods. The damage depends on the access rights of the compromised account. The motives of stealing data might be different from an individual, it might be to take vengeance from an employee or just gain some financial profit. There are different scenarios from which an insider can steal the data drawn from the report [1]. The ways in which are insider can steal data are demonstrated in Fig. 14.1.

Fig. 14.1 Ways in which an insider can steal the data

Insider's Chronicle

From recent studies, it is clear that insider threats are costlier and dangerous than external threats. In this section we will summarize the insider incidents, consequences adduced from surveys and technical reports. In report's [2, 3] states that 68% of the organizations are moderate to extreme vulnerable insider threats and confirms that insider attacks are more frequent than earlier days. A majority of organizations (58%) consider themselves as least effective or even worse in monitoring, detecting and responding to insider threats. They also released the types of insiders that pose biggest risk to an organization categorized in Fig. 14.2.

Insider attacks area unit notably dangerous for 3 reasons:

- Insiders are not malicious all the time but their dangerous activities leads to outside attacks.
- They create weakness in a company or an organization and make their IT security standards vulnerable.
- Insiders expose the sensitive information which leads to the downfall of business reputation.

For these reasons, business executive attacks end in devastating losses for organizations. The outcomes of an insider attack are disclosed in Fig. 14.3.

Paper is categorized as follows: Sect. 14.2 narrates previous works, problem statement and proposed model is stated in Sect. 14.3, results are in Sect. 14.4 and the conclusion is concluded in Sect. 14.5.

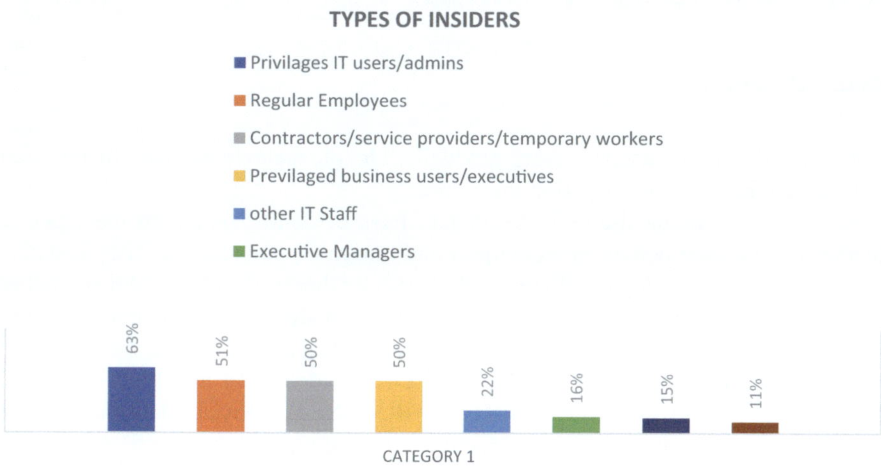

Fig. 14.2 Types of insiders pose the biggest risk to organization

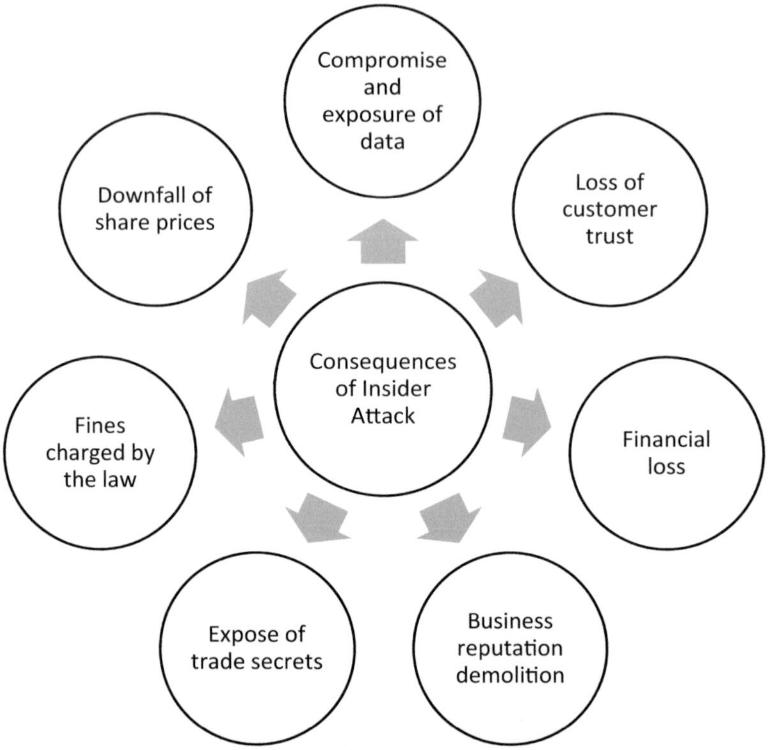

Fig. 14.3 Possible consequences of an insider attack

Related Work

The insider data thefts are increasing gradually, a lot of researchers are doing research and they are helping in mitigation those threats.

In [4], R. A. Alsowail and T. Al-Shehari have evidently explained the types of insider threats, their detection techniques and mitigating techniques. They had also explained the methodology to do research on insider threats to get efficient and better outcomes. Authors had also proposed a 10-question model for a comprehensive study on insider threats with empirical factors. They had discussed about the detection mechanisms like anomaly detection, Misuse-based detection and both combined detection. They also worked on CERT [5], RUU, SEA, ENRON datasets that are used for insider threat approaches. Machine learning algorithms like SVMs, KNN, GMM that are employed in insider threat detection are also hashed.

In [6], an UVM approach method is proposed to prevent the data theft from cloud storage where the authors have developed a data theft prevention system and detection system which will work on the factors what the user is searching and what file is opened. The verification system will work by using two mechanism they are OTP verification and Question verification. This mechanism is for cloud systems and by analyzing their results it shows that will enhance the security.

In [7], an insider attack detection system integrated with packet filtering technique. Here the packets in the intranet are scanned and filtered, if any suspicious packets are found then they are dropped and the admin is alerted. In [8], a Role and User-Based profile assessed insider threat detection is designed in an automated way. Their framework comprises various key parts in which they process the approached log data record and then a client profile is developed along with the current day job. By this the level of threat is evaluated for a person. In [9], a mechanism is proposed to mitigate the insider attacks in the cloud environment. They implemented an alternate methodology to secure the information in the cloud by employing the innovation of hostile imitation. They planned it to guard data acquired in the cloud, recognize anomalous data access designs. If any suspicious unapproved access is detected then it is verified by challenge question. If it is a positive, then a lot of imitation data is sent to the masquerader.

Although, the principal challenge in insider threat region is data theft and to assemble exceptionally exact frameworks that can anticipate, identify and forestall insider assaults consistently. The zone of insider threats and their protection arrangements is as yet not saw quite well. A few overviews attempted to encourage the field by zeroing in solely on hypothetical and reasonable scientific classifications. Our aim is to minimize the insider data thefts by developing a mechanism to detect and monitor the activity continuously. In the following section we have proposed a mechanism to detect and monitor the insider data theft.

Methodology

In this section we illustrate about Descriptive Research Methodology [10] and it is encapsulated in Fig. 14.4.

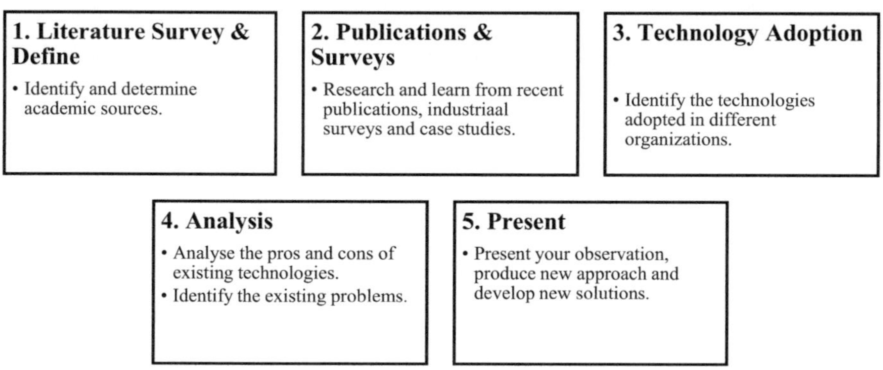

Fig. 14.4 Descriptive research methodology employed in our study

There are five stages in our methodology and they are as follows:

Stage 1:

Literature survey and define: In the first stage, all the theory related work about "Insider threats and thefts" is done to review the existing research. By the help of many search engines we query all the related information by using all the possible search terms (E.g., "Insider data theft detection", "Insider robbery detection", "Insider data breach detection" and "Insider threat detection technologies"). Furthermore, online blogs and related scripts are reviewed for utmost knowledge.

Stage 2:

Publications and surveys: In the second stage, the related publications, research articles and existing surveys are reviewed. The pedagogical sources like Scopus database and IEEEXPLORE are used. Suitable search terms are queried that are specified. During our search all the appropriate synonyms (e.g., cyberattacks, cyberthreats, vectors, issues, data theft, turncloaks, theft detection, etc.) are also reviewed in our study. Remember that not all our retrieved articles are not relevant. So, all the papers must be scrutinized by reading the title, abstract, conclusion and some articles in them. We must include only those articles that have experimental results and empirical evidences. By this step we will finish our theoretical approaches.

Stage 3:

Technology Adoption: In the third stage, we identify the existing technologies that are adopted in an organization. Not only for one organization, we need to collect plethora of data with empirical proofs and performance data. Identification of both the pros and cons of the technology is crucial part. With the help of case studies, we can evaluate the better performing technique used in a company to detect a malicious insider. Almost all the available threat detection techniques are reviewed.

Stage 4:

Analysis: In this stage we analyze all the available technologies with their strengths and weaknesses, and ranked on the basis of their availability. In this study we will identify the reasons and possible solutions to overcome the existing technologies.

Stage 5:

Present: In the final stage, we present a new model with our work on study, analysis and observations. Our new hybrid model will perform better than existing models and it is described in the next sections.

- In our work, we have developed a system with two modules, one is to monitor a directory or a file system continuously and alerts the admin if there are any changes like file creation, modification and deletion.
- The other is to compare the hash values of the new suspicious file with our existing file.
- We have used MD5, SHA1, SHA 256, SHA3_512 and CRC 32 hashing algorithms to generate hash values.

Module 1: Directory Monitoring System

- In directory monitoring system we used a python library called "Watchdog".
- We developed a simple program to monitor directories as specified in the command—line arguments and generates the event logs.
- This will alert the administrator whenever there are changes in a library.
- If any new suspicious file is detected then that is sent to Hash comparison system to compare and verify the file [5].

Module 2: Hash Generator and Verifier

- In this module we used hashlib, zlib libraries to generate MD5 [11], SHA1 [12], SHA256 [13], SHA3_512 [14] hashing algorithms and CRC32 [15] error detecting code.
- We developed a program to generate the hash value of the given file and compare them with the actual key file.
- If the hash values are matched, then it means that the file is the same then definitely they must have stolen that file. In this way we can detect data theft (Fig. 14.5).

- Step 1: Input our Important file.
- Step 2: To generate the hash.
- Step 3: Store and Display the hashes.
- Step 4: If any suspicious file/activity found in our directory then admin is alerted.
- Step 5: The file is sent for hash verification.
- Step 6: If hash values matched then Admin is notified else continue monitoring.

Results and Discussion

Results

Output-1:

Output Explanation:

From Fig. 14.6 we can see that any file creation, modification, deletion is detected and user is alerted. This is directory monitoring system, the given directory is monitored and if any operation takes place in that directory then the user is alerted.

Output-2:

Output Explanation:

In the above case, all the hash values and CRC value matched. So, it is confirm that the file was obtained by stealing it (Fig. 14.7).

Output-3:

Output Explanation:

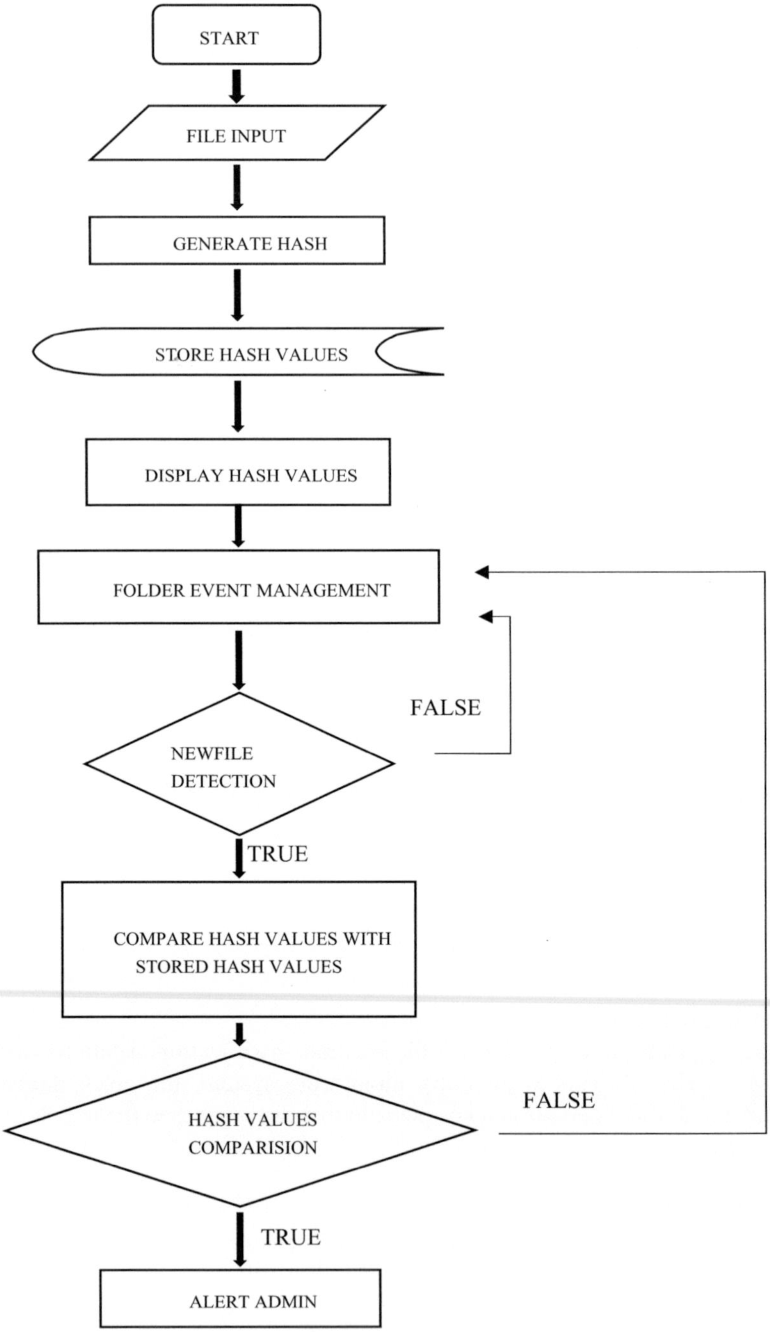

Fig. 14.5 Flowchart of our insider data theft detection and monitoring system

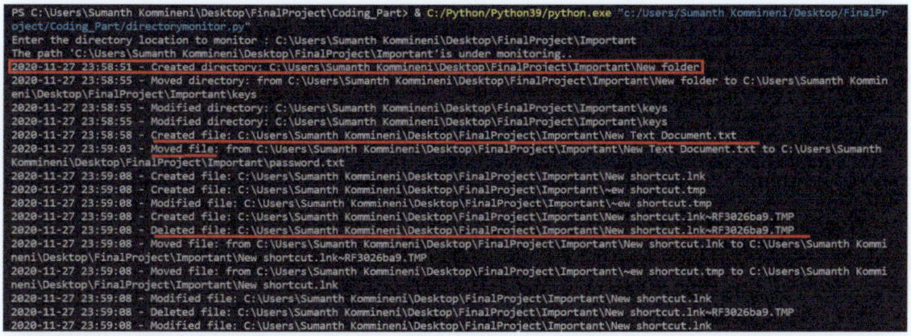

Fig. 14.6 Module 1 Output

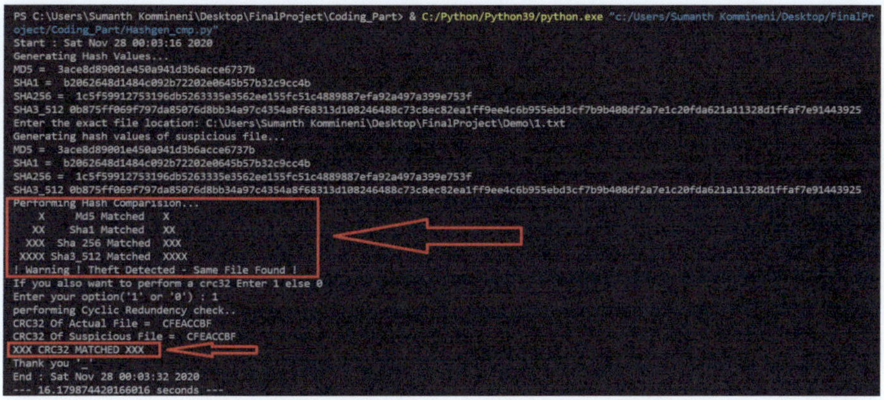

Fig. 14.7 Module 2 output—true case output

In the above case, all the hash values and CRC value did not match. So, it is prompting that the files did not match (Fig. 14.8).

Output-4:

Output Explanation:

As discussed above, our program is designed in a way to detect hash collisions also. When we execute the module 2 code, then the user is asked to provide the suspicious file. In the above case, Only SHA-1 hash value was matching, remaining all did not match, this is called hash collisions. Even if the files are different, in rare cases same hash values are generated to avoid those we are comparing with 4 hash functions. In this way hash collisions are detected (Fig. 14.9).

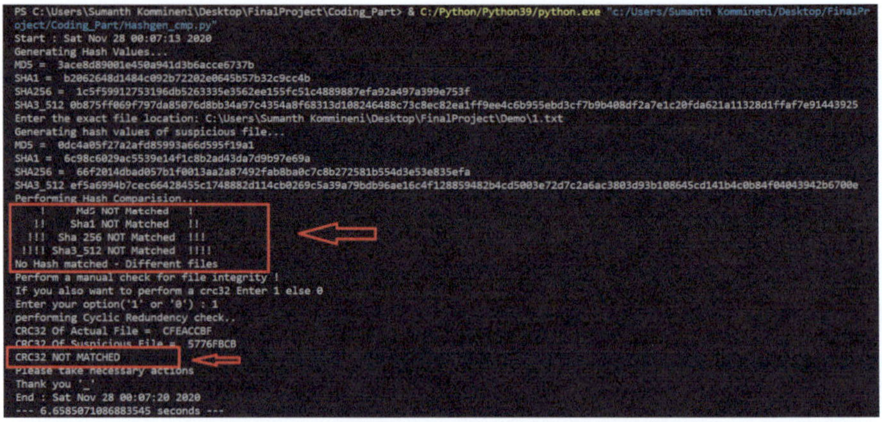

Fig. 14.8 Module 2—false case output

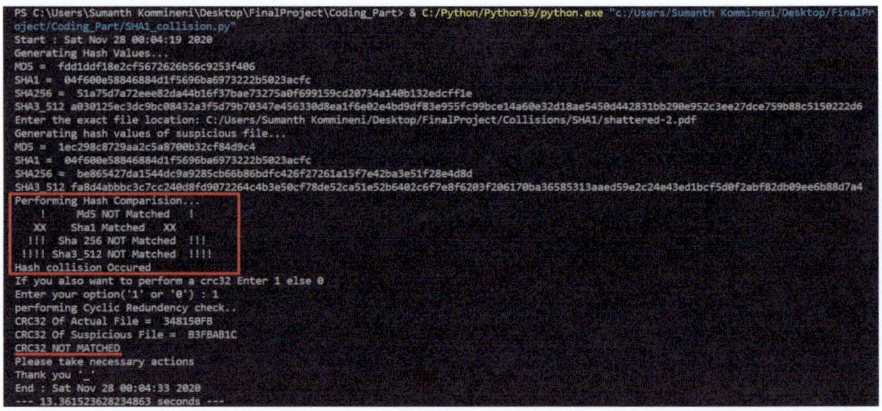

Fig. 14.9 Module 2 output—collision detection

Limitations and Drawbacks

- The main reason of choosing of 5 Hash algorithms is to improve security and integrity.
- We know hash collisions [16] occur in MD5, SHA1 but the chance of occurrence is almost negligence.
- By using 5 algorithms we can detect any theft until the files are not tampered.
- The integrity can be achieved by always by encrypting the file or with password protection, Data/File must be protected with password or encryption mechanism to avoid data tampering.
- It is mandatory to know what data can be stolen.

- All the servers must be located in DMZ only, firewall must be designed in a way not to send any important file out of the intranet.
- User-Based and Role-Based Profile Assessment should be performed.
- Deploy data loss prevention mechanism and data recovery mechanism.
- Only secured and standard protocols must be used in the data transfer process.
- Employees having access to confidential data should be moved to an isolated area to perform any tasks and have to be monitored round the clock.
- A token-based password system must be implemented, so that every time they login a new session and password is generated. This improves authenticity.
- Reduce the use of applications like G-Mail, Dropbox, Social Media which are 3rd party and potentially vulnerable to our system.
- All the tasks performed must be logged, in IDS, NIDS, HIDS false positives must be examined properly.
- Hash comparison is not completely reliable system, there might be discrepancies. So, also perform other file theft and insider theft detection mechanisms.
- Insider threats are always costly, so every small mistake costs a penny. So, operate everything wisely.

Conclusion

Data theft is becoming a pivotal theft in IT Industry. Not only in IT, many institutions, organizations, cloud platform are facing it. In our research, we have analyzed the insider threats and insider data thefts, how serious they are. We have demonstrated a mechanism to monitor and detect the insider data theft. Our program will scan the files, generate the hash values and compare them. It is always a good thing to be secured all the time, attacks may not only happen through outside but also inside. The most important difference between an insider and outside attacker is about the awareness about the system which they will be working with in an organization. An insider can detour the security infrastructures like IDS and raise an attack vector and exploit from any corner which is frightening. So, all employee's must be given with proper guidance and explanation about insider threats, its causes and affect's.

We must be up to date in terms of security to avoid any kind of data breaches and hacks. Further research can be done on our project and project greater and efficient outcomes. Developments can be done, P—Packet Filtering [7] R—Role-Based and User-Based profile assessment [8] A—Anomaly Detection [17] and F—File Access Logs [18], PRAF methodology can be implemented this to obtain greater accuracy and security.

I conclude our research and our knowledge is helpful in terms of minimizing the insider data theft and threats.

References

1. Insider Data Theft: Definition, Common Scenarios, and Prevention, Tips. Available: https://www.ekransystem.com/en/blog/insider-data-theft-definition
2. 2020 Insider Threat Survey Report by GURUCUL, Available: https://gurucul.com/2020-insider-threat-survey-report
3. 2020 Cost of Insider Threats: Global Report, Available: https://www.observeit.com/2020costofinsiderthreat/
4. Alsowail, R.A., Al-Shehari, T.: Empirical detection techniques of insider threat incidents. IEEE Access **8**, 78385–78402 (2020). https://doi.org/10.1109/ACCESS.2020.2989739
5. CERT: Insider Threat Test Dataset. Software Engineering Institute, Carnegie Mellon University. Accessed: May 6, 2018 (2016). [Online]. Available: https://resources.sei.cmu.edu/library/asset-view.cfm?assetid=508099
6. Patel, N., Kansara, K.B.:UBM—UVM approach for preventing insider data theft from cloud storage. In: 2018 5th International Symposium on Emerging Trends and Technologies in Libraries and Information Services (ETTLIS), Noida, pp. 36–40 (2018).https://doi.org/10.1109/ETTLIS.2018.8485217
7. Kim, J.: Development of integrated insider attack detection system using intelligent packet filtering. In: 2011 First ACIS/JNU International Conference on Computers, Networks, Systems and Industrial Engineering, Jeju Island, pp. 65–69 (2011). https://doi.org/10.1109/CNSI.2011.4
8. Legg, P.A., Buckley, O., Goldsmith, M., Creese, S.: Automated insider threat detection system using user and role-based profile assessment. IEEE Syst. J. **11**(2), 503–512 (2017). https://doi.org/10.1109/JSYST.2015.243844
9. Kowsik , R., Vignesh, L.: Mitigating insider data theft attacks in the cloud. In: 2016 Second International Conference on Science Technology Engineering and Management (ICONSTEM), Chennai, pp. 561–567 (2016). https://doi.org/10.1109/ICONSTEM.2016.7560956
10. Descriptive Research: Definition, Characteristics, Methods, Examples and Advantages. Available: https://www.questionpro.com/blog/descriptive-research/
11. Rivest, R.: Step 4. Process Message in 16-Word Blocks. The MD5 Message-Digest Algorithm. IETF. p. 5 (April 1992), sec. 3.4. https://doi.org/10.17487/RFC1321. RFC 1321. Retrieved 10 October 2018
12. Eastlake, D., Jones, P.E.: US secure hash algorithm 1 (SHA1). RFC **3174**, 1–22 (2001)
13. Yadav, B.C., Merugu, S., Jain, K.: Error assessment of fundamental matrix parameters. Lecture Notes Electr. Eng. **500**, 151–160 (2019)
14. Selvakumar, A.L., Ganadhas, C.S.: The evaluation report of SHA-256 crypt analysis hash function. In: 2009 International Conference on Communication Software and Networks, Macau, pp. 588–592 (2009). https://doi.org/10.1109/ICCSN.2009.50
15. SHA3-512, Federal Inf. Process. Stds. (NIST FIPS)–202. https://doi.org/10.6028/NIST.FIPS.202
16. CRC32, Chen, C., Tian, R., Yao, Z.: Design and implementation of the CRC-32 checksum of parallel algorithm. Yi Qi Yi Biao Xue Bao/Chinese J. Sci. Instrum. **34**, 151–155 (2013)
17. Tao, X., Liu, F., Feng, D.: Fast Collision Attack on MD5 (PDF) (2013)
18. Kim, M., Kim, K., Lee, H.: Development trend of insider anomaly detection system. In: 2018 20th International Conference on Advanced Communication Technology (ICACT), Chuncheon-si Gangwon-do, Korea (South), pp. 373–376 (2018). https://doi.org/10.23919/ICACT.2018.8323762

Chapter 15
Automatic Vehicle Alert and Accident Detection System Based on Cloud Using IoT

Pachipala Yellamma, P. G. Sandeep, R. Revanth Sai, S. Rohith Reddy, and D. Mahesh

Introduction

The Internet of thing is a fast-growing number of physical devices (which have a unique identifier) interconnected with some devices over the Internet. These rapid growing physical devices usage has been constantly increasing due to its efficiency, user-friendly nature and simple control which are automatically save time and money [1]. These physical devices are combining and transmit information and transfer data via connected devices exclusive of any person communication [2]. If collision is detected in humongous transportation vehicle, the victim's location and information are sent to concerned officials more quickly. If IoT is implemented in transportation vehicles for pre-programmed accident detection and transmission of victim's location and information to concerned officials, then the loss of life in accidents may gradually decrease [3]. This proposed work is on IoT using Raspberry Pi which is connected with GPS, electronic health record reader and sensors. Initially, people's emergency contacts are recorded in cloud computing and recovered in an incident of the accident. Amazon Web Services (AWS) are the proposed work of the cloud provider we used (AWS). This paper discusses Raspberry Pi as a currently exploring way to provide victims accident-prone location to the emergency contacts given by the victim through Internet, providing a decrease in time consumption to identify accidents and begin treatment to the victims.

The proposed work supports the constant follow up of the vehicle by means a GPS sensor and in the incident of any accidents [4]. The specific id of electronic record of the victim is already stored in a text format in Raspberry Pi. The Internet-based Raspberry Pi transmits vehicle location and a file may provide victims' with medical and emergency contact information to nearby hospitals in the form of an electronic

P. Yellamma (✉) · P. G. Sandeep · R. Revanth Sai · S. Rohith Reddy · D. Mahesh
Koneru Lakshmaiah Education Foundation, Guntur, Andhra Pradesh, India
e-mail: pachipala.yamuna@gmail.com

V. Garcia Diaz and G. J. Rincón Aponte (eds.), *Confidential Computing*,
Advanced Technologies and Societal Change,
https://doi.org/10.1007/978-981-19-3045-4_15

mail service. When the mail is delivered to the authorized personnel, they notify the ambulance and provide the location of accident. Then, the authorized personnel will get back essential data from the cloud and process it to the specific details. Finally, the staff will send SMS on the incidence of the accident with the hospital address to the emergency contact of the victim.

Literature Review

Pachipala Y. proposed a scheme which detects the accident using a piezoelectric sensor which does not support truly static measurements and also not provide exact value when exposed to elevated temperatures [5]. Pham Hoang Oat proposed a system which depends on ARM 11 controller, which is very expenditure [6]. Performance of ARM microcontroller depends on an implementation, and if the programmer does not fix the bugs properly and compile the program properly, then it takes an exceptionally long time to work sufficiently. Pachipala Yellamma, produced a system in which they used a global mobile system; in some cases, there may be no network in some hilly regions; in those cases, the system could not make a reasonable effort [7]. Mishap proposed a system which contains a vibration sensor which is extremely sensitive to high frequency noises [8]. This segment ignores comparable existing arrangements, analyzes their benefits and disservices approach to accident detection; reporting and navigation is an existing system for accident detection, reporting and vehicle navigation. Pachipala Yellamma proposed existing system used shock sensors, NFC tags and GPS to identify the state of the motor vehicle. This method utilizes existed data of the user to identify the persons [9]. Upon the accident happening, deployment of airbags would trigger the shock sensor and the position of the medium would be immediately sent to the server using the HTTP request. Upon receiving request, help can be sent immediately to the location and perform appropriate medical support as person has already been identified.

Jiang Meng proposed an automatic accident detector and reporting system (AADRS) is an advanced and ideologically different system aiming at the road safety [10]. The primary goal of the proposed work is to assist and treat the victims of the road accident [11]. The process behind this system is by automating the alert system by automatically requesting for nearby medical assistance upon a collision or an accident to the vehicles [12]. The task is accomplished by usage of the application by public the incident and by application linked to sensors of the system [13]. Both these entities contribute in determining the location of accident-prone area and provide medical assistance as quickly as possible.

The proposed model is specifying the user alerting system in favor of vehicle accident detection system which is another similar system aimed at providing information about the accident. This system uses vibration sensor to detect abnormal vibrations the vehicle has undergone and reports the geographical location of the vehicle via SMS to the enrolled members associated with the person. Auto Accident App, developed by Platinum.

Proposed Methodology

The proposed model, main idea is to reduce time gathering all the medical information of the victim and location of the accident-prone area, by automatically intimating emergency contacts given by the victim. This is achieved by successful connection of global positioning system (GPS) and accelerometer sensor to Raspberry Pi 4 models B, and the model is associated with Amazon Web Services (AWS). In AWS, we should connect Raspberry Pi 4 model B to AWS IoT hub. The proposed work is use the Raspberry Pi 4 models B with Raspberry Pi an operating system image loaded in Micro-SD-card. Every input operation is done by keyboard and the mouse is attached with the Raspberry Pi, and all the resulting information are displayed in the monitor connected to HDMI cable. After the hardware connections, software connections should be started using the connection we should connect Raspberry Pi 4 model B to AWS IOT core, and after successful connection of AWS IoT hub, all the data gathered by the Raspberry Pi is broadcast to the AWS IoT hub using MQTT broker. All the data is received by the AWS IoT hub and is stored in DynamoDB by making a rule in IoT CORE. The related sensors are continuously sending the measured value when an accident occurs and abnormal values are sent to the AWS IoT. While an abnormal value has been detected in lambda function, it is send to SNS, and the SNS information is forward to the specified emergency contacts. Lambda acts as manager when to send data to SNS and always store the data received sensors in AWS IoT subscribes topic in stored in the DynamoDB. Every attribute and data is present in JSON format. When SNS receives the statistics, immediately it will transmit the location information and victim's information to predefined emergency number. The proposed work contains both hardware and software components; they are explained below.

Raspberry Pi: Raspberry Pi which is in size of a credit card occupies very less space in the automobiles. Raspberry Pi is a uni board computer which can be associated with abundant devices for many purposes and functions similar to an immense one. There are various models of Raspberry Pi m; the model utilized in our proposed work is Raspberry Pi model 3B. This model is designed with 1 GB RAM with 1.2 GHz quad core multiprocessor.

Global Positioning System (GPS): The global positioning system works by using the help of satellite. It is known as satellite navigation system. Everything which has GPS connects to the satellite for reading the raw data of location with timestamp in all weather conditions. It is done within the fraction of the second. This GPS module receives the satellites raw data and transfers to/converts into human understandable format. This transformed information is sent to the emergency contacts via electronic mail. The GPS module is associated with the Raspberry Pi utilizing female 2 female jumper wires and continuous tracking information of the vehicle is sent to the Raspberry Pi, and this information is passed to the cloud.

MPU6050: The MPU6050 sensor module is an incorporated 6-pivot Motion GPS beacon. It has a three-axis gyroscope, 3-hub accelerometer, digital motion processor, and a temperature sensor, all in a solitary IC. It can acknowledge contributions from different sensors like 3-pivot magnetometer or pressing factor sensor utilizing its auxiliary I2C transport. In the event that outer three-axis magnetometer is associated, it can give total 9-hub motion fusion yield. A microcontroller can speak with this module utilizing I2C correspondence convention. Different boundaries can be found by perusing esteems from addresses of specific registers utilizing I2C correspondence. Whirligig and accelerometer perusing along X, Y and Z tomahawks are accessible in 2's supplement structure. Gyrator readings are in degrees each second (DPS) unit; accelerometer readings are in g unit. The MPU6050 sensor is combined with the GPS module; the predetermined GPS module is starts moving as the sensor notice any difficult situation in the vehicle development. At first, the sensor is offered the consistent worth, and when any sort of unusual developments is distinguished, then the steady worth changes and an alternate worth will be noticeable. At the point when an unusual worth springs up, then a mishap happened.

Cloud Storage: Cloud storage is a model of computer data storage in which the data looted from devices is saved and stored in the cloud in a pool of resources said to be cloud storage. The physical storage is owned and managed by a hosting company. In this paper, we use Amazon Web Services called as AWS. All the information obtained from Raspberry Pi is stored in this cloud storage only. It also contains all the listed emergency contacts.

AWS IOT Core: AWS IoT core is a cloud service managed by AWS which enables the users to connect IOT devices easily and securely interact with various cloud applications and other services provided by the AWS. In this, we connect our Raspberry Pi to AWS IOT core. Raspberry Pi is connected to the AWS by utilizing this cloud service.

AWS Lambda: AWS lambda is but a server less computing platform. This was introduced by the Amazon. It will act as a manager; when a data is being received it processes the data relating to the user and sends to the endpoint.

AWS DynamoDB: DynamoDB is an Amazon web administration data set framework that supports data constructions and key esteemed cloud administrations. It permits clients the advantage of auto scaling in memory reserving, reinforcement and re-establishes alternatives for all their web scale applications [15].

AWS SNS: Amazon simple notification service (Amazon SNS) is a managed service that provides messages delivery from sender to recipients. Using this Amazon service, we send electronic mail to victim's emergency contacts. In case of accident, every emergency contact given by the victim will get the alert by SNS [16].

Figure 15.1 shows how to connect MPU6050 to Raspberry Pi, Raspberry Pi model 4 models contains 40 GOIP pins and using female-to-female jumper wires. The proposed system is connecting to the MPU6050 sensor and to the pins in Raspberry Pi. Each bit of the information is broadcasted toward the Raspberry Pi using female-to-female wires.

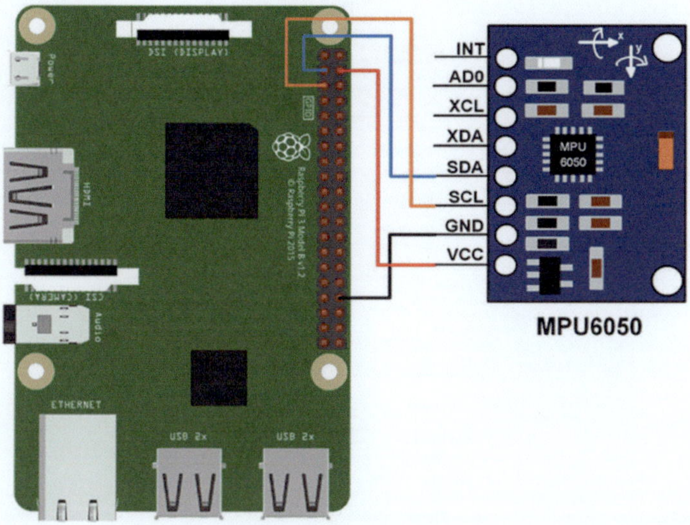

Fig. 15.1 MPU6050 interfacing with Raspberry Pi 4

Figure 15.2 shows how to make a connection between Raspberry Pi and GPS module. The Raspberry Pi 4 contains 40 GOIP pins. The female-to-female jumper wires are connecting with GPS module to Raspberry Pi. All the data is transferred to the Raspberry Pi via female-to-female wires. Fig. 15.3 represents the step-by-step work flow. The proposed algorithm is initial MPU6050 which is a sensor which reads the gyro axis in *x*, *y*, *z* and send the information to Raspberry Pi. Similarly, the same applies to the GPS module; it is also connected with Raspberry Pi using female-to-female jumper wires. The GPS module continuously monitors the vehicle location

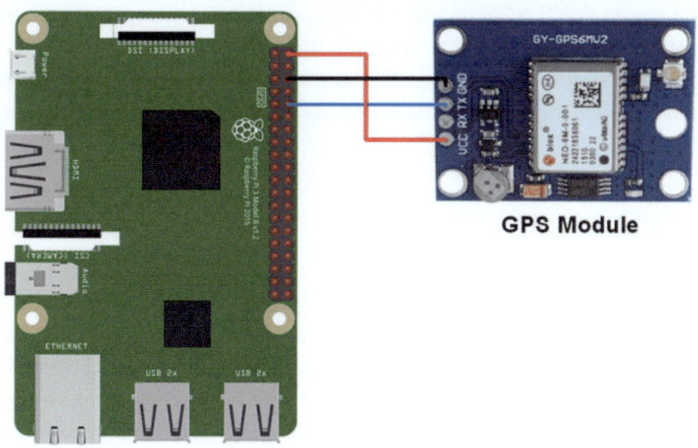

Fig. 15.2 Interfacing of the Raspberry Pi 4 with GPS module

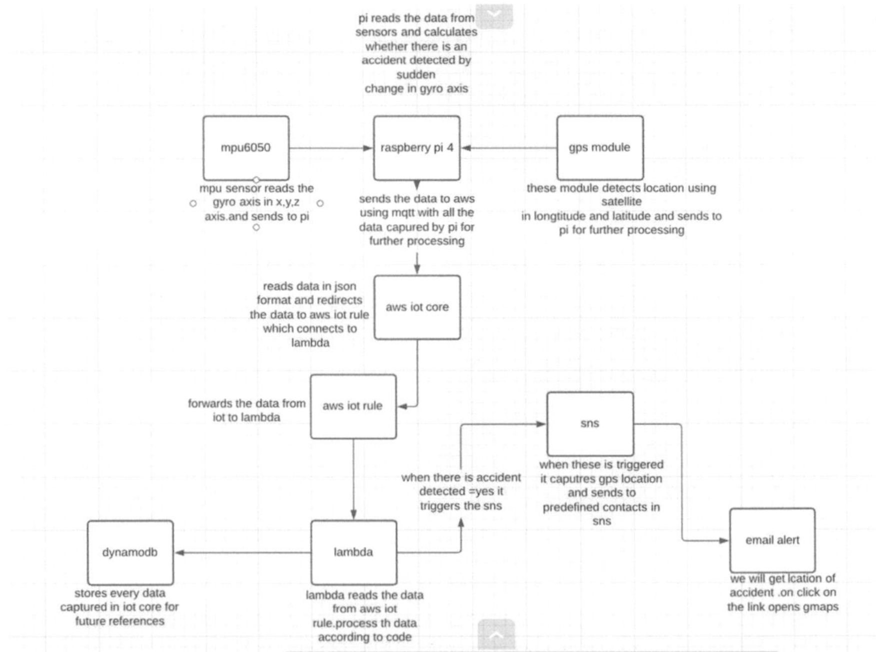

Fig. 15.3 Accident detection and sending alert algorithm

and sends the data to the Raspberry Pi. The Raspberry Pi is connected with the AWS IOT core using Internet. Each bit of the data is broadcasted to the AWS IoT core, and it can read the data in JavaScript Object Notation (JSON) format. Meanwhile, the data redirects to AWS rule which created initially and connected to AWS lambda. Lambda reads the data, and if an abnormal value occurs while monitoring the data, then lambda opens AWS SNS module in this component the spot of accident-prone area. Finally, the medical information sent to the injured party which is attached to an electronic mail and send to the emergency contacts given initially by the victim. If no abnormal value is detected, then lambda forwards all the data to store in DynamoDB.

Experimental Results

After successful implementation of proposed system when an accident occurs, all the emergency contacts would get the following email with accident-prone area location and medical information of the victim. The connecting is from Raspberry Pi to AWS IOT core, transferring data to AWS IOT using MQTT broker. Data is stored in JSON format. Time, location; detection is transferred. AWS IOT core receives data from Raspberry Pi 4 and triggers AWS IOT rule which sends capture data to lambda.

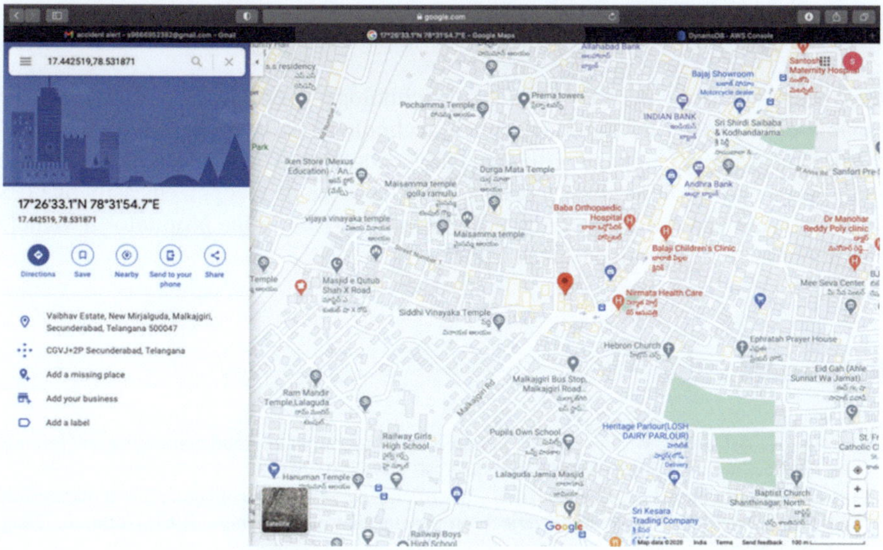

Fig. 15.4 Red pin the GPS location of the accident detected

Dynamo database is tag along with the NoSQL rule. Data is stored JSON format; data is representing in the string format and storing in DynamoDB. Electronic mail received by an emergency contact with location of accident-prone area. This mail triggered by AWS lambda to AWSSNS.SNS sends the email to admin email account.

GPS system is recognizing the location of the accident area, and it specifies red spot in the location. It confirms the accurate spot accident detected as shown in Fig. 15.4. By using proposed system, it is very easy to identify the accident location.

Conclusion

The chance of trailing a life in today's situation is very higher. In this proposed method, real-time lets the authoritative people are willing to retrieve the location of the accident spot immediately supply the medical treatment to the injured party in very little time. This purposed system also deals with various issues which include security and authentication of information. With proper implementation of this, proposed work will preserve the best life-changing technology. We are trying to make the earth a safer place to live. The proposed system identifies and sends the location of the accident spot area immediately to all the victims' emergency contact numbers and send the location to the required local authorities to deploy the resources. We can hope there can be less loss of life in accidents.

Future Scope

This proposed work preserves the best technology ever, also the knowledge be capable of expanded by introducing new technology built in directly by the vehicles manufacturing company with some advancements. The proposed work will be useful for a variety of purposes like as vehicle theft can be recognized easily by GPS sensor present and different security alerts can be given to the proprietor if vehicle crosses certain characterized speed limits.

References

1. Botta, A., deDonato, W., Persico, V., Pescape, A.: Integration of cloud computing and internet of things a survey. Future Gener. Comput. Syst. 684–700 (2015)
2. Hernandez-Ramos, J.L., Bernabe, J.B., Skarmeta, A.: Army: architecture for a secure and privacy-aware lifecycle of smart objects in the internet of my things. IEEE Commun. Mag. 28–36 (2016)
3. Yellamma, P., Anupama, P., Lakshmibhavani, K., Jhansi, U., Priya, S., Kazalalitha, C.: Implementation of e-voting system using block chain technology. J. Crit. Rev. **7**(6), 865–870 (2020). ISSN: 2394-5125
4. Wan, M.: Control cloud-data access privilege and anonymity with fully-anonymous attribute-based-encryption. IEEE Trans. Inf. Forensics Secur. 1–10 (2015)
5. Tiwari, M., Garg, H., Tiwari, R.K., Gupta, S., Yadav, A.K., Deep, A., Jha, M.: Execution of accident vehicle tracking system and a keen application based following framework. IEEE Trans. Inf. Forensics Secur. (2017)
6. Yellamma, P., Abhinav, B., Jaya Vaishnavi, B., Ushaswini, G., Srinivas, M.: Forecasting techniques for sales prediction. Int. J. Adv. Sci. Technol. **29**(6), 3042–3049 (2020)
7. Oat, P.H., Drieberg, M., Cuong, N.C.: Advancement of vehicle tracking system utilizing GPS and GSM modem. IEEE Access (2013)
8. Yellamma, P., Rajesh, P.S.S., Pradeep, V.V.S.M., Manishankar, Y.B.: Privacy preserving biometric authentication and identification in cloud computing. Int. J. Adv. Sci. Technol. **29**(6), 3087–3096 (2020). ISSN: 2005-4238 IJAST
9. Mishap: Accident detection and reporting system utilizing GPS, GPRS and GSM technology. In: IEEE—International Conference on Informatics, Electronics & Vision (ICIEV), pp. 850–859 (2012)
10. Narayana, V.A., Premchand, P., Govardhan, A.: A novel and efficient approach for near duplicate page detection in web crawling. In: 2009 IEEE International Advance Computing Conference (2009)
11. Meng, J.: Plan of vehicle situating system based on ARM. Wen Department of Physics and Electronic Information Engineering
12. Pachipala, Y., Srinivas Rao, T., Siva Nageswara Rao, G., Baburao, D.: An IoT based automatic accident detection and tracking system for emergency services. J. Adv. Res. Dyn. Control Syst. **12**(02), 111–117 (2020). https://doi.org/10.5373/JARDCS/V12I2/S202010013
13. Swetha, K., Shaik, A.S., Usha, S.: Smart dustbin. J. Appl. Sci. Comput. (JASC) **5**(9), 351–359 (2018). ISSN: 1076-5131

14. Sharma, S.K., Jain, K., Suresh, M.: Quantitative evaluation of panorama softwares. In: Kumar, A., Mozar, S. (eds.) ICCCE 2018. ICCCE 2018. Lecture Notes in Electrical Engineering, vol. 500. Springer, Singapore (2019). ISSN: 1876-1100
15. Challa, M.L., Soujanya, K.L.S.: Secured smart mobile app for smart home environment. Mater. Today Proceedings **37**(Part2), 2109–2113 (2020)
16. Ansari, M.D., Usman, M.: Editorial. Int. J. Sensors Wireless Commun. Control **10**(4), 438–439 (2020)

Chapter 16
A Novel Architecture for Detecting and Preventing Network Intrusions

Challa Madhavi Latha, Mohammad Mutayeeb Risalath Ahmed, K. L. S. Soujanya, and D. V. Lalitha Parameswari

Introduction

Information technology has largely improved the performance and quality in modern life (IT). Users can now select from a wide range of devices to meet their requirements, such as high computer processing speeds and fast networks. Unfortunately, as we become more reliant on technology, the number of security incidents rises. Risks, as well as strikes, add the theft of information that is confidential coming from a laptop computer or maybe community server be probably the most vulnerable info saved for a protection intelligence service (SIS). Furthermore, online hackers are able to intercept users' charge card info as well as snoop on the online transactions of theirs, even worse, or, safety–critical products could be jeopardized. Due to the intricacy of risks plus strikes, protection method setup is now more challenging. The contemporary hacking threads are definitely the major danger and issues within the IT business. For instance, online hackers utilize breakthroughs within personal computer processors as well as community rates of speed to boost just how much as well as the pace of malicious site traffic which might end up in DoS or even distributed DoS episode.

As a result, network protection has evolved into an industry devoted to the development of software and hardware platforms for detecting and preventing network threats. One of the most well-known information security principles is the defense-in-depth strategy, which employs a multilayered structural architecture that includes firewalls, vulnerability assessment tools (anti-viruses and worms), and encryption.

C. M. Latha (✉) · M. M. R. Ahmed · K. L. S. Soujanya
Department of Computer Science and Engineering, CMR College of Engineering and Technology, Kandlakoya, Hyderabad, India
e-mail: saidatta2009@gmail.com

D. V. Lalitha Parameswari
Department of CSE, G. Narayanamma Institute of Technology and Science for Women, Hyderabad, India

V. Garcia Diaz and G. J. Rincón Aponte (eds.), *Confidential Computing*, Advanced Technologies and Societal Change,
https://doi.org/10.1007/978-981-19-3045-4_16

The network intrusion detection and prevention system (NIDPS) was designed to be the network's last line of defense. When in detection mode, NIDPS monitors network traffic for any suspicious or unsettling behavior and generates warnings; when in prevention mode, it generates block packet alerts [1].

Literature Survey

As stated by [2, 3], cloud computing has emerged as a real pattern of business IT services clothes airers which provide scalable and cost-effective processing. Meanwhile, software defined networking (SDN) is increasing traction inside business networks because of the power of it to offer better community managing freedom while simultaneously decreasing running expenses. Generally, there seems to be a pattern with the two solutions to collaborate so as to supply the requirements of a business [4]. The brand new problems presented through the matrimony of SDN and cloud computing, especially the ramifications of theirs for business system safety measures, however, are, inadequately comprehended. This particular newspaper seeks to deal with this particular essential concern. We start by thinking about the protection ramifications, particularly the impact on DDoS episode safeguards systems, within a business system which utilizes each solutions. We learned that, in case the protection structure is appropriately set up, SDN technological innovation could possibly aid businesses within protecting against DDoS strikes [5]. To that particular conclusion, we suggest a DDoS mitigation structure which fuses extremely programmable community overseeing for hit detection having a flexible command building for precise and rapid strike effect. To cope with the dataset change problem as well as adjust on the brand new structure, we suggest a graphic model-based encounter detection framework. As per simulation benefits, the structure of ours could efficiently and effectively solve the protection challenges presented through the contemporary community paradigm, as well as the hit detection system of ours may efficiently state a variety of risks utilizing real-world community visitors [6].

Cloud computing will be here to remain, as well as software defined network (SDN) is increasing traction. With equal solutions within the horizon as prospective business IT fixes, it is really worth thinking about the ramifications of merging the two, especially of terminology of business system safety measures [7, 8]. This particular paper investigates the effect of SDN in addition to cloud storage space on DDoS episode safeguard. In accordance with the analysis, we determine the problems as well as advantages presented by these appearing innovative developments. We feel that, with very careful design and style, SDN is able to assist avoid DDoS strikes [9, 10]. In order to help the results, we proposed the DaMask structure as a defense against DDoS strikes. In comparison with various other fixes, DaMask demands the very least quantity of hard work out of the cloud provider, implying in which not many modifications are needed. The SDN-based community keeping track of as well as balance mechanism allows companies to manage as well as personalize the cloud safeguards systems of theirs without interfering with some other cloud people [11].

DaMask [12] additionally comes with a graphical model-based anomaly detection component. We proposed a unit upgrade strategy that makes use of Bayesian inference to upgrade the inference type on a frequent schedule to enhance detection reliability. In order to assess the functionality, we additionally performed a simulation learn utilizing true community traces. The end result shows that the proposed DaMask works well with the unique obstacles. SDN-based community managing could easily adjust to modifications in deep community topology. The detection algorithm prints fast adequate being utilized for Internet package inference and contain a top detection pace. When compared to regenerating applications, the recommended airer upgrade procedure preserves a considerable quantity of period while leading to very little functionality damage of terminology of detection reliability [13, 14].

Some of the main features that attract various attackers are the cloud's open and distributed design, the Internet's vulnerability, and the numerous weaknesses of cloud service models. One of the cloud computing security concerns is a denial of service attack. As a result of this attack, legitimate users' requests are not processed. To protect a network from such sophisticated attacks, a defense mechanism is required. Intrusion detection is primarily used to identify and log attacks. An intrusion detection system is suggested based on a knowledge multithreaded device. A single technique is insufficient to detect such attacks. A multithreaded knowledge-based intrusion detection system (IDS) has been proposed to detect DOS attacks [4].

In order to reduce cyber-attacks, a proposal for an intrusion detection scheme for DoS attacks in the cloud has been made. NIDS should be implemented at the infrastructure layer. To summarize, the goal is to mitigate the impact of a dos attack by detecting it early and accurately so that appropriate countermeasures can be implemented. NIDS will detect the events based on the rules, and this will be notified. Monitoring device resources and file systems, detection using rule-based algorithms, and fuzzy logic to assess the intensity of the attack are some methods for detecting DoS attacks [15]. This approach is combined with others to form a solid solution. The fuzzy model can be modified by changing signatures, and it can be used to detect known attacks effectively.

Personal computer protection is a key matter of the present-day planet. Personal computer networking protection is focused on protecting against unauthorized owners via increasing permission to access a laptop system. An intrusion detection product (IDS) is something that analyses network site visitors as well as operator behavior to differentiate in between non-hostile and hostile site traffic. The vast majority of present-day networks make use of misuse detection or even anomaly detection methods for intrusion detection. Unfamiliar intrusions are not possible to identify utilizing misuse-based IDS, along with anomaly-based IDS possess a top false positive detection fee. So as to deal with this particularly, the suggested method engages a hybrid car intrusion detection as well as avoidance process which fuses networking as well as host-based IDPs [16].

By using just one gadget, administrators can readily monitor the perimeter of a business system inside most three modes, public, namely private, along with wireless networks. By using a host-based phone system, we are able to keep track of each and every multitude inside a personal community for malicious pastime, of course, if the malicious task is recognized, we are able to do something or even change the guidelines from a centralized server process. By using a network-based technique, we are able to keep track of the system for malicious pastime, of course, if the malicious task is recognized, the administrator is able to capture a variety of countermeasures. A wireless-based unit might be utilized to observe as well as stay away from WiFi-based strikes for a business's wireless community.

This particular newspaper details a novel software applications development which use the quality of service (QoS), as well as parallel remedies of Cisco Catalyst, switches to increase the analytical performance of a method intrusion detection in addition to protection system (NIDPS) inside high-speed networks. We have established a real society to indicate assessments creating a Snort NIDPS. Our assessments indicate weak points in NIDPSs, similar to the disappointment of theirs to point to accomplish a number of packets also the productivity to reduced packets to come down with visitors that is high, high-speed networks without examining them. Within Snort's assessment, we counted the number of packets shipped, evaluated, lost, screened, injected and excellent. We recommend making use of QoS put in place answers within a Cisco Catalyst 3560 Series Switch together with parallel Snorts to improve NIDPS efficiency in addition to reduce the variety of fallen packets. The ultimate outcomes of ours show that the novel put in place of ours betters the effectiveness [5, 6, 14].

Attacks on networks have risen sharply in recent years. The main explanation for this is that people with insufficient training have unrestricted access to and usage of software (written and posted to websites by technical experts). Several types of guided attacks can cause network disruptions on purpose. These attacks target the application layer and other layers of the TCP/IP protocol set. The internal body can also attack the network, in addition to the external body. An IDPS, on the other hand, is widely regarded as one of the most effective technologies for detecting threats and assaults. These attacks target the application layer and other layers of the TCP/IP protocol set. The internal body can also attack the network, in addition to the external body. An IDPS, on the other hand, is widely regarded as one of the most effective technologies for detecting threats and assaults. To boost NIDPS analysis efficiency and reduce NIDPS processing time, we recommend using a QoS configuration and parallel technology. As a consequence of our approach, devices can be designed to make it easier to the attacks [15].

Software defined networking (SDN) plus balance principle is important part of Network Implementation of Function Virtualization (NFV). We determine utilized situation for NFV control by using SDN in addition to the controlling principle with this demo. In order to manage virtual network function (VNF) cases along with the GENI testbed, we utilize RINA's managing structure (a clean slate Recursive InterNetwork Architecture). Snort is an intrusion detection system that we make use of. Source of energy, as well as desired destination hosting companies, several IDSes, a receptive vSwitch (OVS), as well as an OpenFlow controller, are an aspect of the community topology of ours. A sent out managing framework operating on RINA screens the express on the VNF cases as well as directs the information to a proportional integral (PI) controller, which in turn transmits ton balancing information on the OpenFlow controller. The advantages of utilizing a control-theoretic ton balancing strategy and also the RINA managing structure wearing virtualized locations for NFV control are evidenced within this specific demo. Additionally, it implies that the GENI testbed is able to support a multitude of NFV-related and SDN tests [7].

This particular demo shows exactly how management principle could be put on to SDN-based NFV control. We likewise show which RINA's managing structure has a selection of attributes that may be utilized to streamline NFV flexible management. Experimentation happens on the GENI testbed. GENI is a groundbreaking testing center that can cater to numerous types of examinations, such as lengthy affecting SDN and NFV. The VNF is dependent on the widely used Snort IDS, as well as website traffic, which is routed to various Snort IDS situations for processing. The benefits of utilizing a control theoretic ton balancing strategy over a standard round method for ton balancing

Analysis

The systems development life cycle (SDLC) is the method of developing or perhaps changing solutions and the designs and methodologies that people work with to create these systems, in methods engineering, information methods, and software engineering. The SDLC definition can serve as the basis for a number of a program growth methodologies deeply in software engineering.

Pre-existing System

Intrusion attacks impact each community software, and numerous products, like firewalls, rule-based intrusion detection, and signature-based intrusion detection as well as reduction systems, are developed to fight them. This employs a rule-based intrusion detection and prevention system in which packets are shot with WINPCAP and also analyzed with Snort. Snort includes a much faster detection method, although it can slow down if perhaps malicious customers ship many packets in a quite short stretch of time as a result of poor process speed or buffer overflow. Some packets are going to be dropped/ignored by Snort, causing them to be prepared by the server. Most intrusion programs will maintain incoming packets within a buffer, and also, the packets may be dropped due to processor slowness or maybe buffer overflow.

Disadvantages of the existing system

1. Increase the volume and speed of malicious traffic.
2. Security issues.
3. Performance degradation.

Proposed System

The paper demonstrates a brand new concept referred to as parallel processing with queues, in addition to a brand new guideline to identify flooding packets. To fix this method and problem throughout the packets, a novel quality of program (QoS) structure was created to enhance the effectiveness of intrusion detection as well as avoidance. Our analysis suggested as well as evaluated an answer which organized packets/traffic inside a multi-layer switch utilizing a novel QoS setup as well as utilized parallel strategies to raise package processing velocity. The brand new structure was examined consuming a bunch of automobile traffic fees, tasks, and kinds.

Advantages of the proposed system

1. Increased performance
2. Increased processing speed

Implementation

WINPCAP produced packets are used as a result of different servers to employ the process, and then reading through all the packets and keeping them inside queues for further processing. The 'pcap packets' folder is made up of each WINPCAP packets. All packets are stored in binary style of WINPCAP, and then we are able to go through and thing to accomplish them with the Python 'SCAPY' API. All packets are preserved within the 'pcap' data format by WINPCAP.

Details of PCAP file The .pcap file extension is frequently linked to Wireshark, a network evaluation application .pcap documents are knowledge files that are produced with the software and contain community packet data. These documents are mostly used to examine a details set's network characteristics. As they are monitored, the data typically help to efficiently manage visitors on a system. The pcap file extension is utilized to preserve the data as well as results of the network review that is why they are known as .pcap files. These documents are accustomed to be able to assess network status and also to go over data marketing communications using Wireshark. Since Wireshark can be obtained for Windows, Mac OS Linux, and X, these .pcap paperwork can be opened too, given that the necessary applications are fitted. Wireshark, WinDump, tcpdump, Packet Square—Capedit, along with Ethereal are all illustrations of traffic exchanges which can easily open up .pcap data.

Test Approach:

Testing can be done in two ways:

1. Bottom-up approach
2. Top-down approach

Bottom-up Approach:

Starting with the smallest and lowest-level units, testing can be done one at a time. In bottom-up testing, a short program executes each module and provides the necessary data, allowing the module to function as it does when embedded in the larger system. As the lower-level modules are reviewed, the focus shifts to the higher-level modules that use the lower-level modules.

Top-down approach:

This particular means of examining starts within the greatest amount on the component hierarchy. Stubs are crafted since the comprehensive responsibilities which are usually performed in reduced fitness-level regimes are not provided. A stub is a component layer that is known as by way of a higher-level component and also comes back information on the calling component indicating the interaction was profitable. There is simply no make effort to look at the reduced lex's reliability.

Validation:

The system has been successfully tested and implemented, ensuring that all of the specifications specified in the software requirements specification have been met. In the event of incorrect input, error messages are shown.

TEST CASES:

Test-case Id	Test-case-name	Test-case-Desc	Test-steps			Test-case-status	Test-priority
			Step/	Expected/	Actual/		
01	Upload PCAP packets patterns	Verify file available or not	If it is not	There is no process	PCAP packets file loaded	High	High
02	Load packets to queue	Verify packets file uploaded or not	If it is not	We cannot load packets into queue	We can see total packets loaded into queue	High	High
03	Run NIDPS detection parallel mode	Verify packets are arranged in queue or not	If it is not	We cannot run NIDPS	We get packets processing with multiple threads	High	High
04	Malicious packets detection graph	Verify packets processing completed or not	If it is not	We cannot get the graph	We get the count and type of attack graph	High	High

Conclusion

A new NIDPS deployment architecture was developed and tested. Computer networks have recently made significant advances in terms of their ability to support varying speeds and data volumes. As a consequence of the fast development, personal computer networks are actually much weaker than ever before to high-speed episodes as well as risks. These may lead to a few of troubles for personal computer networks. A lot of high-speed strikes are hard to find and stay away from. Examining boosting quantities of website traffic becomes tougher and tougher because of fast technical advances which are rising community speed. To meet the security needs of network systems and users, a variety of open-source resources have recently become available. This paper investigates the success of an open-source project. The evaluation's goal was to see how well the NIDPS did for high-speed visitors when restricted by off-the-shelf hardware, after which discover how you can make it better. This particular analysis concentrated on the weaknesses of that protection method as NIDPS to come down with high-speed community reception. We proposed a repair for this particular flaw, in addition to a brand new structure for NIDPS generation which uses parallel solutions and QoS to coordinate as well as boost community managing as well as targeted traffic processing effectiveness, therefore boosting NIDPS efficiency. As a consequence of the brand new design and style of ours, Snort's productivity enhanced significantly, letting a lot more packets to become examined just before becoming dropped. The effectiveness of Snort NIDPS (analysis, identification, and then avoidance rate) has grown to a lot more compared to ninety-nine %. Snort NIDPS prepared as many as eight Gbps with no interruption that uses two devices (PCs) associated with two one GB interfaces.

This particular quantity could be enhanced to 32 Gbps by putting in additional NIDPS nodes, and that is the optimum unit ahead bandwidth. The study's objective was creating a technical fix which was in theory audio. This particular expertise helps with generalizing the issue as well as remedy, which makes it a lot easier to use the suggested method of infrastructures apart from the examination foundation applied to this specific research.

References

1. Shaik, S., Shaik, A.S.: Design of accessible display using ARM9 to control home area networks. Int. J. Sci. Eng. Technol. Res. (IJSETR) **03**(40), 8046–8050 (2014). ISSN: 2319-8885
2. Wang, B., Zheng, Y., Lou, W., Hou, Y.T.: DDoS attack protection in the era of cloud computing and software-defined networking. Comput. Netw. **81**, 308–319 (2015)
3. Chauhan, K., Prasad, V.: Distributed denial of service (DDoS) attack techniques and prevention on cloud environment. Int. J. Innov. Adv. Comput. Sci. **4**, 210–215 (2015)
4. Samani, M.D., Karamta, M., Bhatia, J., Potdar, M.B.: Intrusion detection system for DoS attack in cloud. Int. J. Appl. Inf. Syst. (Foundation of Computer Science), **10**(5) (2016)

5. Vasudeo, S.H., Patil, P., Kumar, R.V.: IMMIX-intrusion detection and prevention system. In: Proceedings of International Conference on Smart Technology Management Computing, Communication, Controls, Energy Mater. (ICSTM), pp. 96–101 (2015)
6. Bul'ajoul, W., James, A., Pannu, M.: Improving network intrusion detection system performance through quality of service configuration and parallel technology. J. Comput. Syst. Sci. **81**(6), 981–999 (2015)
7. Akhtar, N., Matta, I., Wang, Y.: Managing NFV using SDN and control theory. Dept. CS, Boston Univ., Boston, MA, USA, Tech. Rep. BUCSTR-2015-013 (2015)
8. Prakash, L.N.C.K., Suryanarayana, G., Ansari, M.D., Gunjan, V.K.: Instantaneous approach for evaluating the initial centers in the agricultural databases using k-means clustering algorithm. J. Mobile Multimedia **18**(1), 43–60 (2022)
9. Kenkre, P.S., Pai, A., Colaco, L.: Real time intrusion detection and prevention system. In: Proceedings of 3rd International Conference on Frontiers in Intelligent Computing, Theory and Application (FICTA), pp. 405–411. Springer, Bhubaneswar, India (2015)
10. Narayana, V.A., Premchand, P., Govardhan, A.: A novel and efficient approach for near duplicate page detection in web crawling. In: 2009 IEEE International Advance Computing Conference, IACC 2009 (2009)
11. Merugu, S., Reddy, M.C.S., Goyal, E., Piplani, L.: Text message classification using supervised machine learning algorithms. Lecture Notes in Electrical Engg. **500**, 141–150 (2019)
12. Li, M., Deng, J., Liu, L., Long, Y., Shen, Z.: Evacuation simulation and evaluation of different scenarios based on traffic grid model and high performance computing. Int. Rev. Spatial Planning Sustain. Develop. **3**(3), 4–15 (2015)
13. Kim, J.-M., Kim, A.-Y., Yuk, J.-S., Jung, H.-K.: A study on wireless intrusion prevention system based on snort. Int. J. Softw. Eng. Appl. **9**(2), 1–12 (2015)
14. Vemuri, R.K., Reddy, P.C.S., Puneeth Kumar, B.S., Ravi, J., Sharma, S., Ponnusamy, S.: Deep learning based remote sensing technique for environmental parameter retrieval and data fusion from physical models. Arabian J. Geosci. **14**(13) (2021)
15. Cisco (2016) Cisco Interfaces and Modules, Cisco Security Modules for Security Appliances. Accessed: Feb. 30, 2018. [Online]. Available: http://www.cisco.com/c/en/us/support/interfaces-modules/securitymodules-security-appliances/tsd-products-support-series-home.html
16. Merugu, S., Jain, K., Mittal, A., Raman, B.: Sub-scene target detection and recognition using deep learning convolution neural networks—ICDSMLA 2020. Lecture Notes in Electrical Engineering, pp 1082–1101. Springer, Singapore (2020)

Chapter 17
Cyber-Hacking Breaches for Demonstrating and Forecasting

T. Guru Akhil, Y. Pranay Krishna, Ch. Gangireddy, Anumandla Kiran Kumar, and K. L. Sowjanya

Introduction

While mechanical arrangements can strengthen digital frameworks against assaults, breaks in data proceed to be the major issue. This energizes us to portray the growth of incidents of data breaks. That will not as it were expand our understanding of information that penetrates but will also shed light on various methodologies, such as protection, alleviate the harm. However, the advancement of accurate digital risk measurement to manage the tasks of protection rate is beyond the span of the current understanding of information breaks. Many accept that protection will be useful. The current examination is propelled by a few inquiries that have not been investigated as of recently, for illustration, are data breaks brought almost by advanced attacks growing, reducing, or settling? The principled reaction to this request will grant us an absent from into the common circumstance of advanced dangers. This request was not replied by past examinations. The datasets, which are examined during 2000 to 2008, does not truly contain the break events that brought approximately by advanced attacks; the datasets inspected. Xu et al. [1] are afterward, be that as it may contain two sorts of events: careless breaks (i.e., scenes brought almost by misplaced, arranged of, taken contraptions and different reasons) and vindictive penetrating. We propose a system for anticipating assault rates within the sight of extraordinary qualities. The technique dissects the extraordinary worth wonder by means of two corresponding methodologies: the Time Series Theory and the Extreme Worth Theory. Regardless the reality that our contextual investigation depends on explicit digital assault information gathered by a honeypot, the strategy can be similarly applied to break down

T. Guru Akhil · Y. Pranay Krishna · Ch. Gangireddy · A. K. Kumar (✉)
Department of ECM, Koneru Lakshmaiah Education Foundation, Guntur, Andhra Pradesh, India
e-mail: kiran.anumandla@kluniversity.in

K. L. Sowjanya
Department of CSE, CMR College of Engineering & Technology, Kandlakoya, Telangana, India

V. Garcia Diaz and G. J. Rincón Aponte (eds.), *Confidential Computing*, Advanced Technologies and Societal Change,
https://doi.org/10.1007/978-981-19-3045-4_17

any digital assault information of its sort. In particular, we make two commitments [2].

Previous Work

The current examination is propelled by a number of requests those that were not explored as of not long prior, for case, Are data enters brought approximately by computerized attacks growing, diminishing, or settling? A principled reaction to this request will deliver us an absent from into the common circumstance of computerized perils. This request was not answered by past examinations. The datasets, which are tested from the beginning of 2000 to 2008, does not contain the break events that are brought approximately by advanced ambushes; the dataset broke down is afterward, be that as it may contain two sorts of scenes: careless enters (i.e., events brought approximately by misplaced, arranged of, taken contraptions and distinctive reasons) and harmful entering. Since careless enters talk to more human botches than advanced ambushes, we do not think approximately them within the current examination. Since the vindictive penetrates concentrated, Zhenxin et al. [3] carry four sub-classes: hacking (counting malware), insider, installment cards deception, obscure; this investigation is going to be will zero in on the hacking sub-classification (called a break in hacking datasets from there on), while taking note of that the other three sub-classes are fascinating all alone and ought to be examined personally. As of late, scientists began displaying information break occurrences. Maillart and Sornette examined the factual properties of the individual personality misfortunes within the joined together states, between year 2000 and 2008. They founded sum of break episodes significantly increments from 2000 to July 2006, however stays stable from that point. Edwards et al. broke down the dataset containing 2253 break episodes that length longer than 10 a long time (2005 to 2015). They found that not one or the other the measure nor the recurrence of information penetrates has expanded throughout the long term. Wheatley et al. inspected a dataset that is solidified from relates to progressive enter events. From the years 2000 until 2015, they found the repeat of colossal entrance events (i.e., the ones that break in abundance of 50,000 records) that happened to US firms is free of time; in any case, the repeat of tremendous enter scenes happening to non-US firms appears an extending design [4, 5].

Proposed System

In this paper, we make the accompanying three commitments. To begin with, we show that the hacking penetrates both occurrence entomb appearance times (reflecting episode recurrence) and break sizes ought to be demonstrated by stochastic cycles, instead of by conveyances. Reports. We are finding that a specific point cycle can enough depict the advancement of the hacking penetrate occurrences Between times of appearance and that a specific ARMA-GARCH demonstrate can sufficiently portray the advancement of the hacking break size, where ARMA is abbreviation for "Auto Backward and Moving Normal" and GARCH is abbreviation for "Summed up Auto Regressive Conditional Heteroskedasticity." We appear that the stochastic cycle models can anticipate the between advent time and the penetrate sizes. To the most amazing aspect our insight, this is often the essential papers indicating that the stochastic cycles, as opposed to circulations, ought to be utilized to demonstrate this digital danger about factor [6]. Second, we find a positive dependency on the between the occurrences between appearance time and the break size and appear that this reliance can be enough portrayed by a specific copula. We likewise show that when foreseeing, it takes place between advent time and break sizes, is important to think about the reliance; in any case, the expectation results are not precise [7, 8]. To the most awesome aspect our insight, this is the main work indicating the presence of this reliance and the outcome of overlooking it. Third, we coordinate couples objective and assessable pattern. The digital hacking exams break occurrences [4, 9]. We locate that the circumstance is in fact deteriorating regarding the episodes between appearance time in light of the reality that hacking break occurrences become increasingly incessant, yet the circumstance is balancing out as far as the occurrence penetrate size, showing that the harm of individual hacking break episodes would not deteriorate [4]. We believe the current examination will spur more examinations, which can offer significant bits of information into substitute peril control draws close. Such encounters are important to protections offices, government organizations, and controllers since they have to be significantly comprehend the thought of data break risks [10–12].

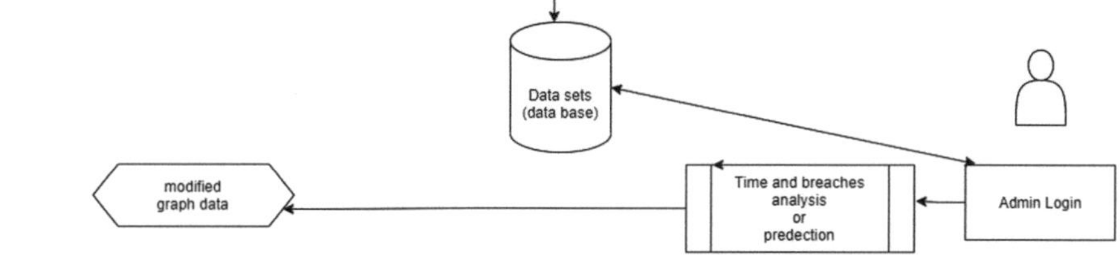

Algorithm

The "Support Vector Machine" (SVM) is the directed AI calculations which can be gained for both order and back slide difficulties. Nonetheless, it is for the most part utilized in course of action issues. In this calculation, we plot every data thing as a point in n-dimensional spaces (where n is number of highlights you've got) with the evaluation of each component being the evaluation of a particular organization [2]. At that point, we perform order by finding the hyper-plane that isolated the 13 two classes well undoubtedly (take a gander at the underneath depiction). Backing vectors are basically the co-ordinates of person recognition. Backing vector machine can be a boondocks which best limits the two classes (hyper-plane/line). In this SVM algorithm, it creates a perpetual dimensional space, a hyper-plane or set of hyper-planes can be utilized for diverse assignments like special case of recognizable proof. The hyper-airplane with the most noteworthy removes to the closest data arrangement objective of any lesson (gathered utilitarian edge) since when all is said in done. The bigger the edge, the lower the classifier speculation botch [12–14]. In spite of the fact that the primary issue can be communicated in a restricted dimensional space, it regularly happens that the sets to isolate are not straight particular in that space. Consequently, the primary confined dimensional space was prescribed be arranged into a parcel higher-dimensional space, which is likely to form the partition simpler in that space [5, 13, 15].

PSEUDOCODE:

```
Characterize number of highlights + 1 as F and SVs + 1 as SV
   FOR each SV
         FOR each feature of SV
      Read streamed data
      Convert it to drift
      Store into array_SVs[SV][F]
         END FOR
   END FOR
   Read streamed data
   Convert it to drift
   store into array_ay[0](b valve)
   FOR each SV
         Read streamed data
         Convert it to drift
         Store into array_ay[SV]
   END FOR
   FOR each highlight
         Read streamed data
      Convert it to float
         Store into array_test[F]
   END FOR
```

```
FOR each feature
        Clear array_AC[F]
END FOR
FOR each SV
    FOR each highlight of the SV
        array_AC[F] + = array_ay[SV]*array_SVs[SV][F]
        END FOR
END FOR
FOR each highlight
        Distance_value + = array_AC[F]*array_test[F]
END FOR
Distance_value- = b
IF(Distance_value > = th)THEN
    RETURN 1
ELSE
    RETURN-1
END IF
```

Module Implementation

Upload Data

The information asset is possible to transfer the data set to both overseer and approved client. The data may be transferred with key to keep up the mystery about the information that is not delivered without information on client [1, 16]. The clients are approved dependent on their subtleties that are shared to administrator and administrator can approve every client. Just authorized clients are authorized to get to the framework and transfer or solicitation for documents.

Access Details

The entrance of information can be gotten from the data set, given by chairmen. Transferred information is overseen by administrator, and administrator is the lone individual to give the rights to deal with the getting to subtleties and favor or unapproved clients dependent on their subtleties [17].

Results

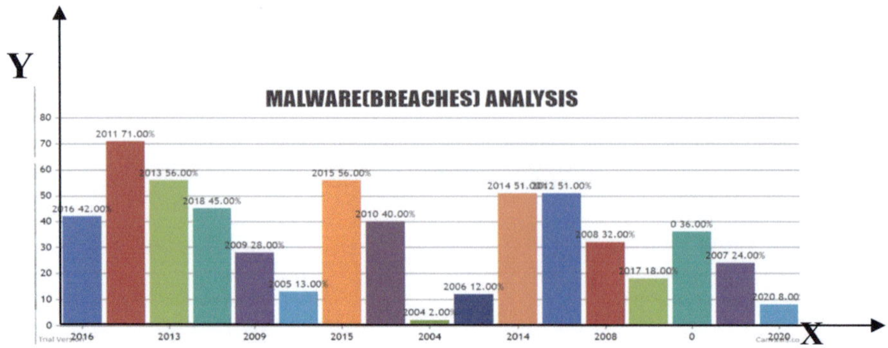

Conclusion

We broke down a hacking penetrate datasets from the points of see of occasions between appearance times and the break estimate and uncovered that the two of them got to be illustrated by stochastic cycles instead of circulations. The factual models created in this paper show acceptable fitting and forecast exactness. Specifically, we propose utilizing a copula-based strategy of managing with expect. The joint likelihood to an occurrence with a particular greatness of break size will happen amid a future time outline. Measurable tests appear that the systems proposed in the paper are superior, to those which are displayed in the composed in light of the truth that the final one is disregarded both the worldly relationships and the dependence between the occurrences between the periods of appearance and the penetrate size. We directed subjective and quantitative investigations to draw further experiences. We drew a bunch of network protection experiences, counting that the danger of digital hacking break occurrences is in reality deteriorating regarding their recurrence, yet not the extent of their harm. The philosophy introduced in this paper can be grasped or balanced to break down datasets of a comparable sort [13].

References

1. Xu, M., Schweitzer, K.M., Bateman, R.M., Xu, S.: Modeling and predicting cyber hacking breaches. IEEE Trans. Inf. Forensics Secur. **13**(11), 2856–2871 (2018)
2. Edward, C., Angela, H., Michel, C.: Analysis of computer security incident data using timeseries models. IEEE **13**(12), 77–86 (2008)

3. Zhenxin, Z., Maochao, X., Shouhuai, X.: Characterizing honeypot-captured cyber attacks: statistical framework and case study. IEEE **8**(11), 1775–1789 (2013)
4. Zhenxin, Z., Maochao, X., Shouhuai, X.: Predicting cyber attack rates with extreme values. IEEE **6**(11), 1–12 (2015)
5. Kishore, P.V.V., Venkatram, N., Sarvya, Ch., Reddy, L.S.S.: Medical image watermarking using RSA encryption in wavelet domain. In: 2014 First International Conference on Networks and Soft Computing (ICNSC2014), pp. 258–262. IEEE (2014)
6. Narayana, V.A., Premchand, P., Govardhan, A.: A novel and efficient approach for near duplicate page detection in web crawling. In: 2009 IEEE International Advance Computing Conference, IACC 2009 (2009)
7. Hogan, J., Adams, N.M.: A study of data fusion for predicting novel activity in enterprise cyber-security. IEEE **3**(11) (2018)
8. Prameela, R., Santi, A., Shaik, A.S.: Tongue controlled wheel chair movement. Int. J. Eng. Sci. Comput. **6**(11), 3272–3274, ISSN: 2321–3361 (2016)
9. Kumar, T., Bajaj, R.K., Ansari, M.D.: On accuracy function and distance measures of interval-valued Pythagorean fuzzy sets with application to decision making. Scientia Iranica **27**(4), 2127–2139 (2020)
10. Fang, Z., Xu, M., Xu, S., Hu, T.: A framework for predicting data breach risk: leveraging dependence to cope with sparsity. IEEE (2021)
11. Das, S., Mukhopadhyay, A., Shukla, G.K.: i-HOPE framework for predicting cyber breaches: a logit approach. IEEE **8**(11) (2013)
12. Devadasu, G., Sushama, M.: A novel multiple fault identification with fast fourier transform analysis. In: 1st International Conference on Emerging Trends in Engineering, Technology and Science, ICETETS (2016)
13. Sapienza, A., Bessi, A., Damodaran, S., Shakarian, P., Kristina.: Early warnings of cyber threats in online discussions. IEEE **4**(8) (2017)
14. Rajeswari, M., Amudhavel, J., Pothula, S., Dhavachelvan, P.: Directed bee colony optimization algorithm to solve the nurse rostering problem. Comput. Intel. Neurosc. (2017)
15. Ahmed, M., Laskar, R.H.: (2019) Eye detection and localization in a facial image based on partial geometric shape of iris and eyelid under practical scenarios. J. Electr. Imag. **28**(3), 18, 033009 (2019)
16. Merugu, S., Reddy, M.C.S., Goyal, E., Piplani, L.: Text message classification using supervised machine learning algorithms. Lecture Notes in Electrical Engg. **500**, 141–150 (2019)
17. Al-Mohannadi, H., Mirza, Q., Namanya, A., Awan, I., Cullen, A., Disso, J.: Cyber-attack modeling analysis techniques: an overview. IEEE **5**(10) (2016)

Chapter 18
Enhanced Security with Crystography Using AES and LSB

Kovuri Harshini, Bandla Naresh, Kutumbaka Sahitya, B. B. V. Satya Vara Prasad, and Batti Tulasi Dasu

Introduction

Cryptography and steganography are well-known for detecting the presence of details. Steganography refers to the process of concealing a message that it would make no sense to someone else but the intended receiver, whereas cryptography refers to the assistance of translating plaintext into an unreadable format. Steganography can also be applied to cryptographic data, so the suggested algorithm improves the security of this data where steganography and cryptography techniques that are really accurate. While there is a probability that combining all approaches straight forward will result in the initial message being intercepted by the attacker, there is also a probability the attacker may intercept the secondary message. Thus, the plan is to incorporate all of them together with more layers of protection and to have a very highly secure data hiding technique. In the wireless network, the field is where this strategy is applied. So, the ultimate intended of algorithm is to create a modern technology that is incredibly protected even though the stego-data is retrieved by attackers [1]. The Advanced Encryption Standard (AES) algorithm is used here, and the encrypted message is then inserted. According to a unique key within an image using the LSB process, it is not incredibly safe to conceal the data using LSB alteration alone.

The suggested approach is to use the LSB technique to provide surveillance with the aid of AES for the confidential information found inside the cover image. This algorithm is used to read the hidden data, and the concealed image is the first step. By the LSB encoder in the concealed picture, the next is to hide the details [2]. The

K. Harshini · B. Naresh · K. Sahitya · B. B. V. Satya Vara Prasad (✉)
Department of ECM, Koneru Lakshmaiah Education Foundation, Guntur, AP, India
e-mail: bbhanuprasad@kluniversity.in

B. T. Dasu
Department of CSE, CMR College of Engineering & Technology, Kandlakoya, Telangana, India

V. Garcia Diaz and G. J. Rincón Aponte (eds.), *Confidential Computing*,
Advanced Technologies and Societal Change,
https://doi.org/10.1007/978-981-19-3045-4_19

stego-image is then sent to AES encryption to compute the attacker's data, where the pixels are encoded. The stego image is recovered using AES decryption, and the output is sent to the LSB decoder, where the secret data is eventually recovered. Consequently, the message is reliably transferred from one end to the other.

Literature Survey

Vikas Singhal proposed the Advanced Secure Hashing algorithm—256 to secure data. The encrypted text is inserted using LSB techniques in image format. The hashed data is modified in a way that makes it completely unreadable in cryptographic hashing in [3]. The 256-bit hash described above will be nearly impossible to transform back to its original state; here, it cannot solve this. The implementation of steganography and cryptography using Advanced Encryption standard algorithm and least significant bit techniques is described in [4]. The encrypted image or text inside a concealed file is indicated by this method. Malathi proposed that steganography and SPIHT compression were implemented using an asymmetric cryptographic algorithm in this paper. Steganography is used with SPIHT compression where image wavelet transformation occurs by AES cryptographic algorithm. Using SPIHT encoding, the cover image is used to compress and encrypt the secret image. To perform decryption over the stego-image in [5], using the Diffie-Hellman technique is to share the key. The file is exposed to SPIHT decompression after decryption which is used to recover the initial data from an image. It is possible to combine steganography with cryptography to improve computer confidentiality. In [6, 7], new approach to provide 24-bit protection is proposed in steganography and cryptography integration. In this method, using the technique randomized method based on LSB is applied to mask and hide the information. This modern optimized approach ensures the improvement in the hiding ability of the information. Then, [8, 9] focuses on the analysis of three methods focused on LSB techniques that means that in each pixel of the image, the bits of message are located in the LSB. In addition, it recommends an improved approach to LSB-based image steganography. The Deflate algorithm is a Lossless' data compression algorithm which incorporates the LZ77. The Huffman algorithm is the length of the concealed message. Protection of the reduced secret data by the AES algorithm is another significant aspect. Bandekar proposed that the primary aim of this paper is to use LSB strategies to conceal hidden data, providing data protection using the AES algorithm. Peak signal-to-noise ratio (PSNR), mean square error (MSE), where the high-quality picture provides a higher PSNR value and estimates algorithm performance. Saini and Verma [10] propose an advanced approach that incorporates both encryption and steganography for image protection. Using the proposed new variant of the AES algorithm, the image is encrypted and then covered in the stego-image using the principle of steganography. This hybrid technique promises higher protection against threats. Audhi and Mascarenhas [11] target on a tactic to assure the safe transfer of data over a network that has modified the mode of data transmission. This is performed using the AES algorithm and steganography techniques for bit plane complexity segmentation. To get a different image, snap the cover image. Finally, to acquire two stego-images, these encrypted images are further inserted separately on the original carrier image [12].

Anwar suggests a combination of LSB and AES Base64 to provide security for messages and diverse file formats contained in digital images. Until being embedded into the picture, using the LSB process and using AES hidden messages are encrypted and Base64 techniques in [13, 14]. The research also proven the performance of the combination of LSB and AES Base64 algorithms on different files and the size of the cover image. An image with RGB channels is used for cover images. Implementation of separable and reversible encrypted knowledge is concealed in encrypted image as a workaround using the AES and lossy technique. AES algorithm, using LSB technique to conceal the cipher data in encrypted image, device automatically generates all three respective keys in [15]. Receiver can work as per the respective keys, such as if he only posses data hiding and image using decryption key, then receiver can only get the image in its original form, or if receiver has data hiding and data decryption key, then can get original data, system also provides auto-generated key protection and auto-generates system mail if the user fails to complete any operation [16].

Working Methodology

Both approaches cryptography and steganography are used to give the data's authenticity to hide from its abuse. The merger of the two results insecure and confidential source of knowledge can be kept secret easily. The key aim of cryptography is solely to mask the message from any insecurity and avoid unauthorized access to it. Essentially, the notion of cryptography and steganography is to have secret and secure communications (Fig. 18.1).

Algorithm for generating stego-image and embedding data into image:

Step a) Start.
Step b) Add the information that needs to be transferred to the receiver.
Step c) Enter specified key of some length.
(The user must enter the password to secure their information).
Step d) Specify the cover image with its path and extension.
Step e) Besides the AES encryption algorithm, the key encryption and plain text are converted into cipher text.
Step f) Using LSB, In the image a steganographic method is embedded with cipher text and a stego-image is obtained.
(This image is communicated through a channel to the recipient.)
Step g) At the receiver's end enter the path and extension of the stegno image.
Step h) Steg analysis is done on the stego image and cipher text is obtained.
Step i) A protected key is acquired from which data can only be accessed from AES decryption by AES decryption algorithm, encrypted text is translated to plain text. (Actual information message received by the sender is obtained).

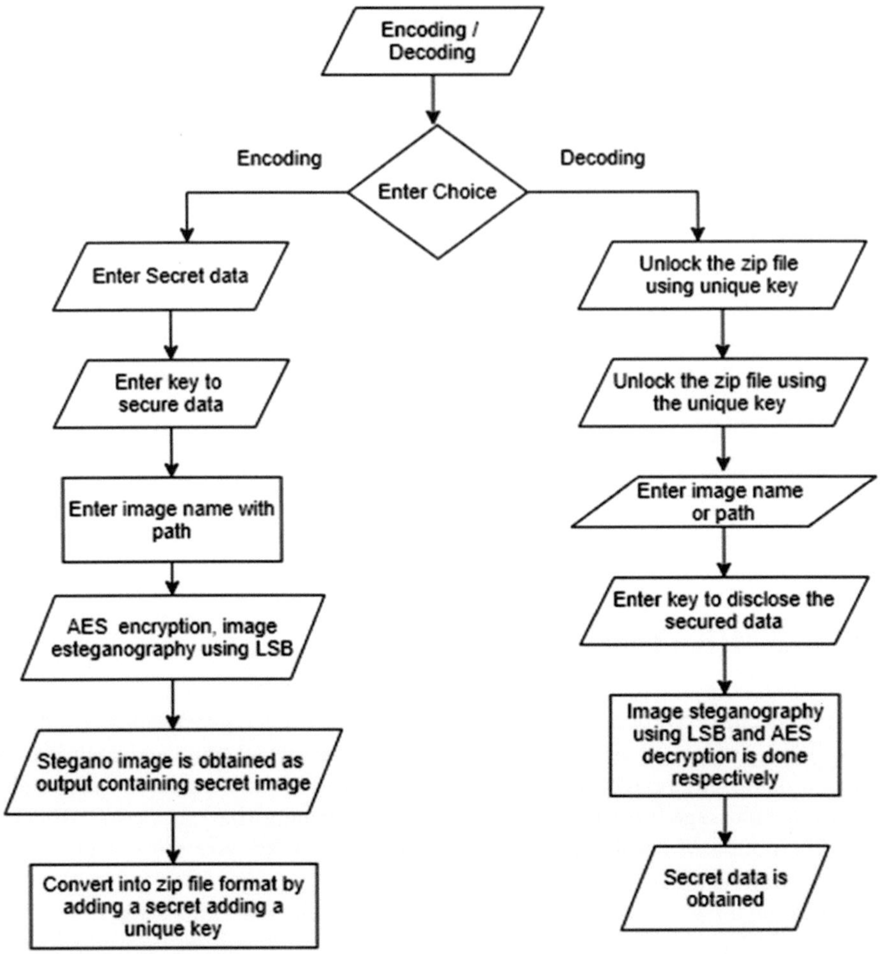

Fig. 18.1 Interface diagram

Step j) End.

Block Diagram

The underlying model of the proposed system is seen in Fig. 18.2. The application of an approach from one end to another is not normally applied in this paper. Here, the sender first enters the message and then applies the AES encryption methods that now have a specific key, which in turn transforms it into cipher text. In addition, the translated encrypted text is inserted using LSB techniques to the concealed image. Finally, the information transmission system is set, and this is called in simple terms as

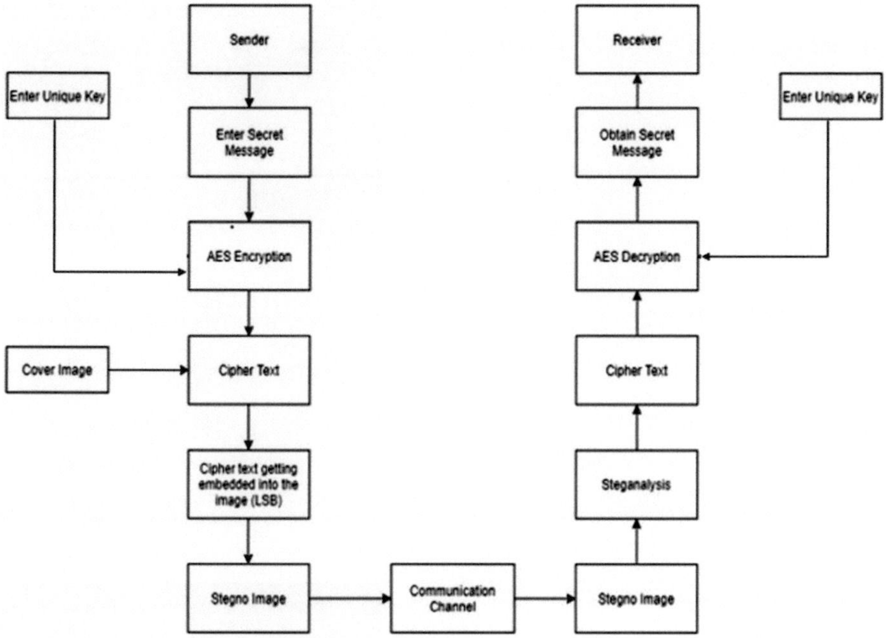

Fig. 18.2 AES combined with LSB from sender's to receiver's end

image with steganography or stego-image sent via a zip file. The sent stego picture is decrypted via a channel to the recipients by applying the AES decryption techniques; additionally, the specific key is shared to the sender. The data is eventually decrypted from the cipher text after imposing all the techniques.

Histogram analysis:

A histogram of image is a schematic view of the pixel distribution strength in a visual image. For each strength value, it charts the number of pixels. Figure 18.7 shows the red, green, and blue cover plane histograms, stego, and encrypted image. It reveals that the concealed image and stego-image histograms are the same. Therefore, it does not support stego analysis. The histogram for image of the cover and the encrypted image is entirely different and provides no sign of the primary image. It therefore avoids statistical attack.

Results and Discussion

It is believed that AES has good protection for block cipher key sizes, something that in many tasks is very useful applications. Using key sizes, they can proceed a quicker execution. Timings for the schemes, which are useful for frameworks where

Fig. 18.3 **a** Secret image, **b** cover image 1, **c** cover image2

Fig. 18.3 (continued)

Fig. 18.3 (continued)

Fig. 18.4 **a** Stego-image1, **b** stego-image2

schemes are used as a crucial factor is real-time efficiency (Figs. 18.3, 18.4, 18.5, 18.6, 18.7, and 18.8).

Figure 18.8 shows that a red, green, and blue plane histogram of the stego image and the retrieved image. This discloses that two histograms are precisely the same. Therefore, this approach guarantees hidden image recovery without failure.

Fig. 18.4 (continued)

Fig. 18.5 **a** Encrypted stego1, **b** encrypted stego2

Fig. 18.5 (continued)

Fig. 18.6 Secret image

Conclusion

For better results, it provides satisfying variables by integrating the characteristics of cryptography and steganography. Creation, preparation, and implementation of the AES algorithm and LSB strategy. It uses the AES algorithm to encrypt the results encoded using the shape of the LSB to mask confidential data and steganography. Throughout the process, the data is still shielded. Transmission of an available channel is over the network. The proposed structure offers great complexity and

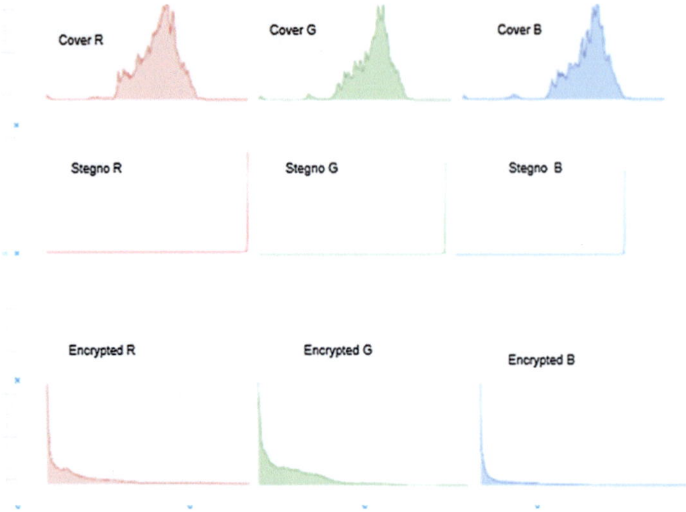

Fig. 18.7 Histogram for cover, stego

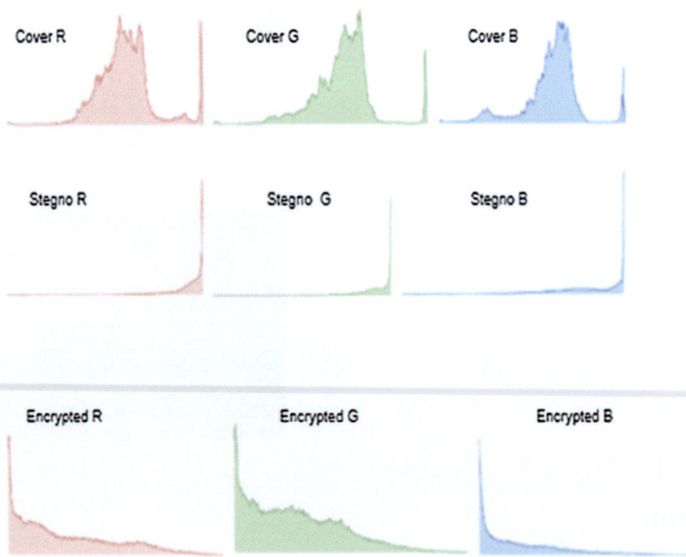

Fig. 18.8 Histogram for secret and encrypted image and recovered image

security. Therefore, this strategy is beneficial to hackers when you encounter trouble accessing the secret data.

References

1. Soujanya, K.L.S.: Ontology based variability management for dynamic reconfiguration of software product lines. J. Adv. Res. Dyn. Control Syst. **9**(Special Issue 18), 2361–2375 (2018)
2. Narayana, V.A., Chamakura, A., Gandi, R.: Deceptive call recognition in a network using machine learning. Acta Technica CSAV (Ceskoslovensk Akademie Ved), **63**(6), 909–914 (2018)
3. Singhal, V., Kumar Shukla, Y., Prakash, N.: Image steganography embedded with advance encryption standard (AES) securing with SHA-256. Int. J. Innov. Technol. Expl. Eng. (IJITEE) (2020)
4. Hybrid approach to text and image steganography using AES and LSB technique. Int. Res. J. Eng. Technol. (IRJET) (2018)
5. Malathi, M., Rahul, M., Kumar, N.S., Thamaraiselvan, R.: Enhanced image steganography using AES and SPIHT compression. In: 2017 international conference on innovations in information embedded and communication systems (ICIIECS) (2017)
6. Phadte, R.S., & Dhanaraj, R.: Enhanced blend of image steganography and cryptography. In: 2017 International Conference on Computing Methodologies and Communication (ICCMC) (2017)
7. Aruna Suhasini Devi, Y.: Ranking based classification in hyperspectral images. J. Eng. Appl. Sci. **13**(7), 1606–1612 (2018)
8. Chikouche, S.L., Chikouche, N.: An improved approach for lsb-based image steganography using AES algorithm. In: 2017 5th International Conference on Electrical Engineering—Boumerdes (ICEE-B) (2017)
9. Bandekar, P.P., Suguna, G.C.: LSB based text and image steganography using AES algorithm. In: 2018 3rd International Conference on Communication and Electronics Systems (ICCES) (2018)
10. Saini, J.K., Verma, H.K.: A hybrid approach for image security by combining encryption and steganography. In: 2013 IEEE Second International Conference on Image Information Processing (ICIIP-2013) (2013)
11. Audhi, S., Mascarenhas, M.: Secure mechanism for communication using image steganography. In: 2019 2nd International Conference on Intelligent Computing, Instrumentation and Control Technologies (ICICICT) (2019)
12. Shukla, A., Merugu, S., Jain, K.: A technical review on image super-resolution techniques,—advances in cybernetics, cognition, and machine. In: 2020, Lecture Notes in Electrical Engineering, Springer, Singapore, pp. 543–565 (2020)
13. Anwar, F., Rachmawanto, E.H., Atika Sari, C., Ignatius Moses Setiadi, D.R.: StegoCrypt scheme using LSB-AES base64. In: 2019 International Conference on Information and Communications Technology (ICOIACT) (2019)
14. Devadasu, G., Sushama, M.: A novel multiple fault identification with fast fourier transform analysis. In: 1st International Conference on Emerging Trends in Engineering, Technology and Science, ICETETS (2016)
15. Kadam, P., Nawale, M., Kandhare, A., Patil, M.: Separable reversible encrypted data hiding in encrypted image using AES Algorithm and Lossy technique. In: 2013 International Conference on Pattern Recognition, Informatics and Mobile Engineering (2013)
16. Gali, N., Venkateshwar Rao, B., Shaik, A.S.: Color and texture features for image indexing and retrieval. Int. J. Electron. Commun. Comput. Eng. **3**(1), 10–14, ISSN: 2249–071X (2012)

Chapter 19
A Smart Security Systems Using National Instruments myRIO

S. Krishnaveni, M. Harsha Priya, S. Mallesh, and Ch. Narendar

Introduction

A smart bank security system is significant in cities during cut-off time. The key purpose of smart bank security system is security and safety. Innovative security features are possible with NI-myRIO using LabVIEW software. Laboratory Virtual Instrument Engineering Workbench (LabVIEW) is a graphical programming language that uses icons instead of lines of text to create applications.

The camera monitors surroundings continuously, and in emergency case, it will communicate with myRIO device. When IR sensor recognizes any movements in premises, then the camera records 30 s video and sends it to officials. The code is implemented with the help of LabVIEW software by taking a case structure, local variable, global variable, and sequence structure.

National Instrument's myRIO-1900 is a portable reconfigurable I/O device that can be used to design control, robotics, and mechatronics systems. The NI-myRIO-1900 is an efficient portable board used for real-time evaluation to enhance the application efficiency. The NI-myRIO-1900 provides analysis of audio signal that includes analog input, analog output, digital input, and digital output in a compact embedded device. The NI-myRIO connects to a host computer over USB and wireless 802.11b, g, n. The MXP breadboard provides a 300-tie breadboard and 50-tie bus bar on an expansion card compatible with the myRIO expansion port (MXP) [1–6]. The MXP breadboard functions as an I/O breakout, with digital and analog signals mapped to two headers on either side of the breadboard itself. The MXP breadboard's connector is also keyed to ensure correct connection with NI-myRIO (Fig. 19.1).

S. Krishnaveni (✉) · M. Harsha Priya
CMR College of Engineering and Technology, Hyderabad, India
e-mail: krishnaveni@cmrcet.ac.in

S. Mallesh · Ch. Narendar
CMR Technical Campus, Hyderabad, India

V. Garcia Diaz and G. J. Rincón Aponte (eds.), *Confidential Computing*,
Advanced Technologies and Societal Change,
https://doi.org/10.1007/978-981-19-3045-4_20

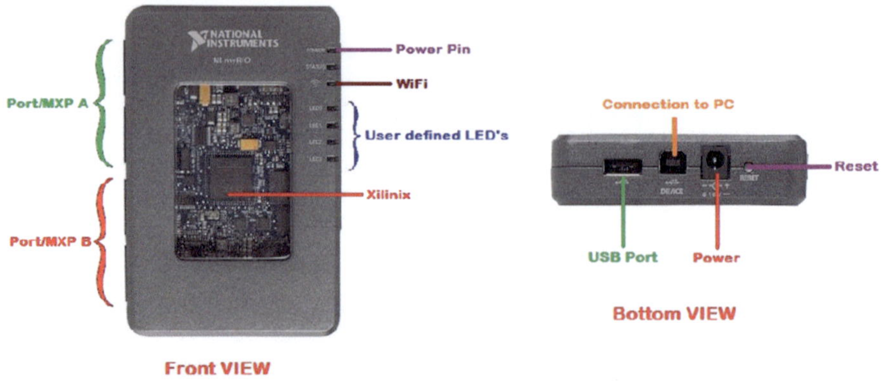

Fig. 19.1 myRIO device

Technical Approach

NI-myRIO-1900 notifies the movements in restricted region with infrared sensor if any abnormalities are observed, and a small recorded video of intruder is sent to the officials in real time using SMTP through mail. USB cameras are picture-capturing cameras; here, the technology used is USB 3.0 which transfers the captured data [2, 3, 7, 8].

The system NI-myRIO uses enhanced image analysis techniques, and hence, the captured images and recorded video are efficiently processed through wireless network data managing system. The system acquires images using a web camera, and continuous audio and video monitoring is also recorded. If any uncertain odd movements are detected from infrared sensor, then it will trigger the buzzer. Further, the processor starts alerting the officials through mail by sending the recorded video along with audio using SMTP in real time.

Methodology

Above shown Fig. 19.2 block diagram describes the implementation of smart security system and interfacing of components.

NI-myRIO is an embedded evaluation board used for developing applications in real time; in this, NI-myRIO buzzer and infrared sensor are connected externally. NI-myRIO is configured with LabVIEW software through PC or laptop. The processor is also integrated with USB camera. Whenever infrared sensor detects any movements, the buzzer is turned on, the camera starts recording the 30 s footage, and immediately, the footage is sent to registered email account [8–14].

In Fig. 19.3, case structure and loops are used to implement the functionality of sensor. The case structure executes only one case at a time. When infrared sensor

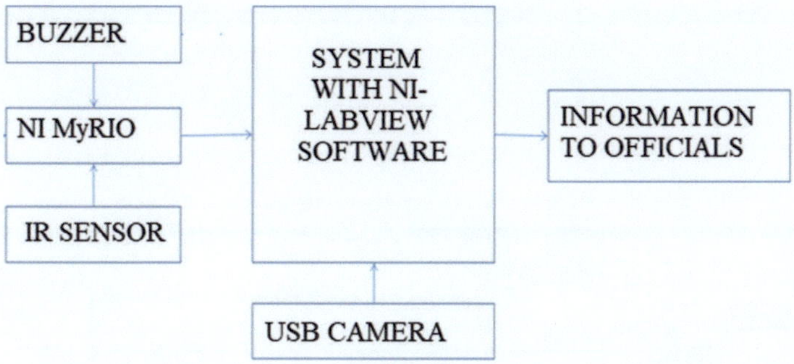

Fig. 19.2 Block diagram for security system

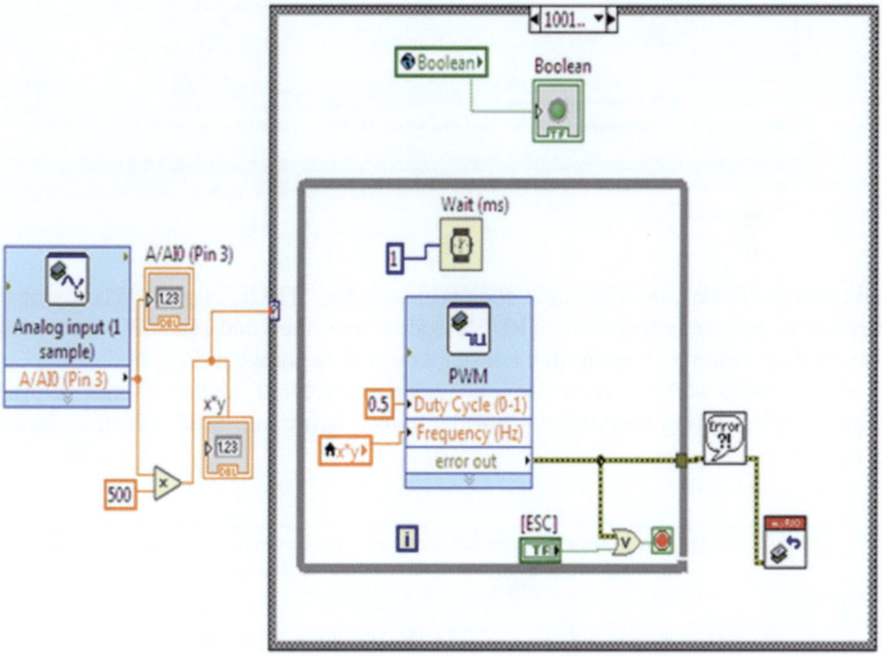

Fig. 19.3 Infrared sensor alerts the processor

detects the obstacles, then it will run the sensor code which alerts the buzzer by sending "TRUE" in the local variable.

When there is no intrusion, code runs with "False" case structure which means there is no odd movements detected by giving "normal" condition in string indicator. Infrared sensor indicator acts as local variable to alert the buzzer. Buzzer indicator acts as global variables to trigger the next devices [15, 16].

In Fig. 19.4, camera captures the surroundings if buzzer indicator is "TRUE". The captured video is stored with an extension of ".avi". If global variable of buzzer

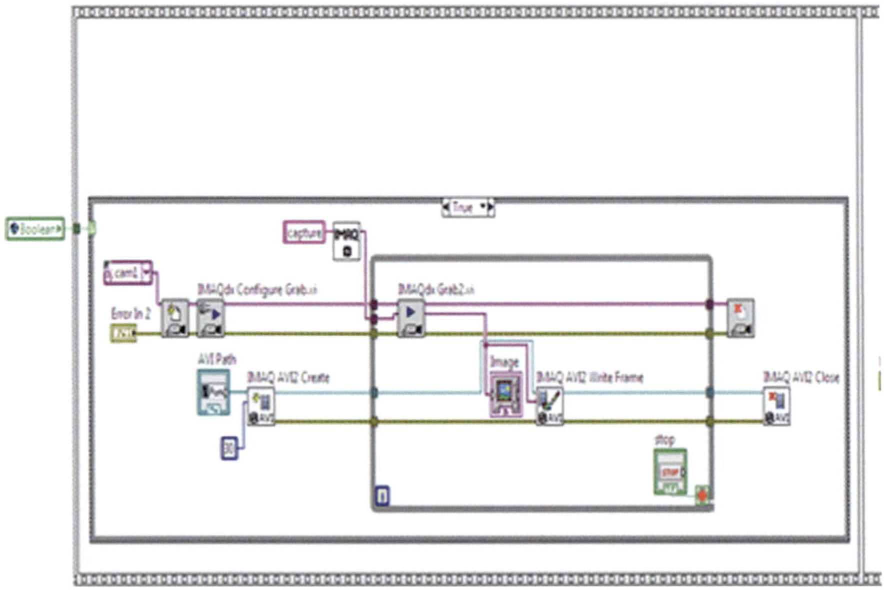

Fig. 19.4 Video capturing

indicator is "False", then the next code will wait for "TRUE" signal. When global variable of buzzer indicator is "TRUE", code is executed and sends 30 s captured surrounding video with audio to the official mail id as shown in Fig. 19.5.

If the global variable is "False", no video is captured, and mail code returns "False" by indicating everything is normal in the "string indicator" which is shown in Fig. 19.6.

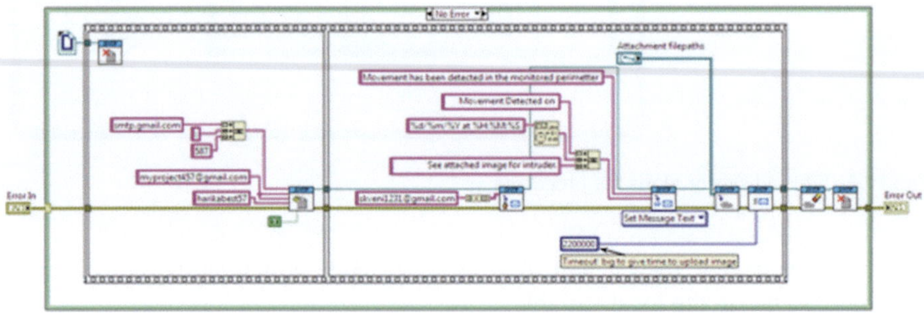

Fig. 19.5 Informing to officials

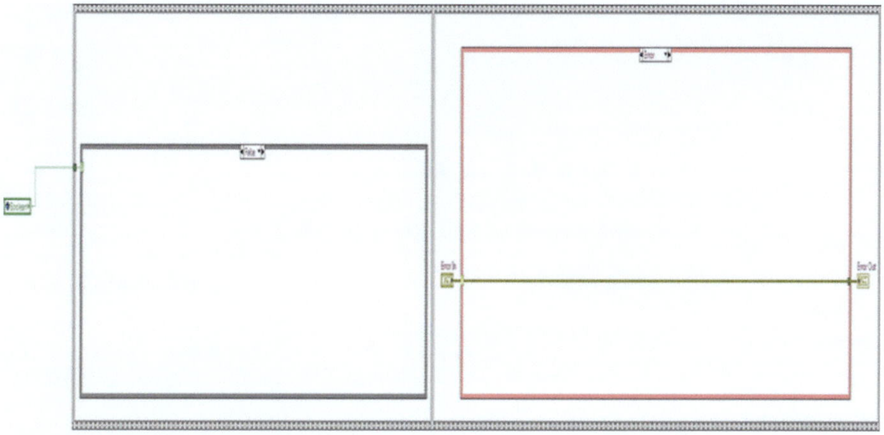

Fig. 19.6 Global variable "False" case

Results

Figure 19.7 shows the hardware setup for smart security system for banks. When infrared sensor senses any object, the case structure will run the "1001 and above case" to give a "TRUE" indication to alarm. The alarm alerts the camera by sending a TRUE value to camera code, then the camera starts capturing the 30 s video with audio. The captured video is saved in ".avi" form in a given location. The 30-s footage is sent to officials with the help of NI-myRIO portable device.

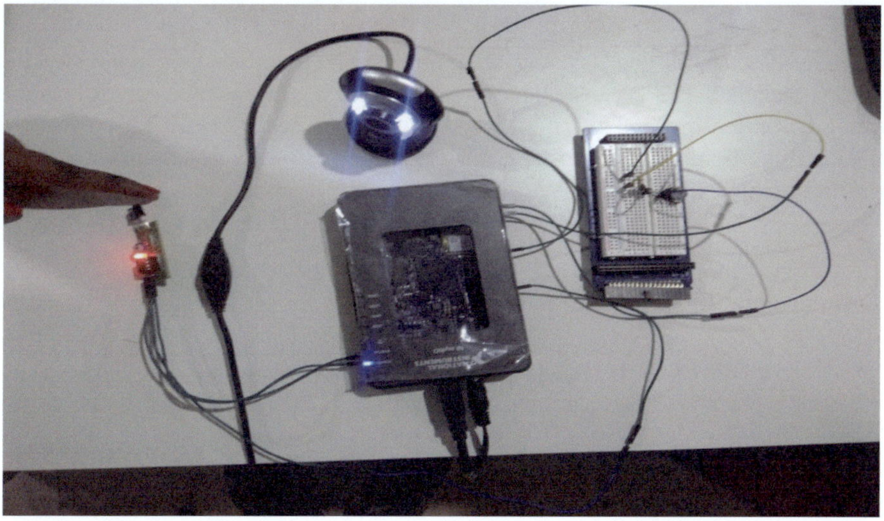

Fig. 19.7 Hardware setup of the project while detection

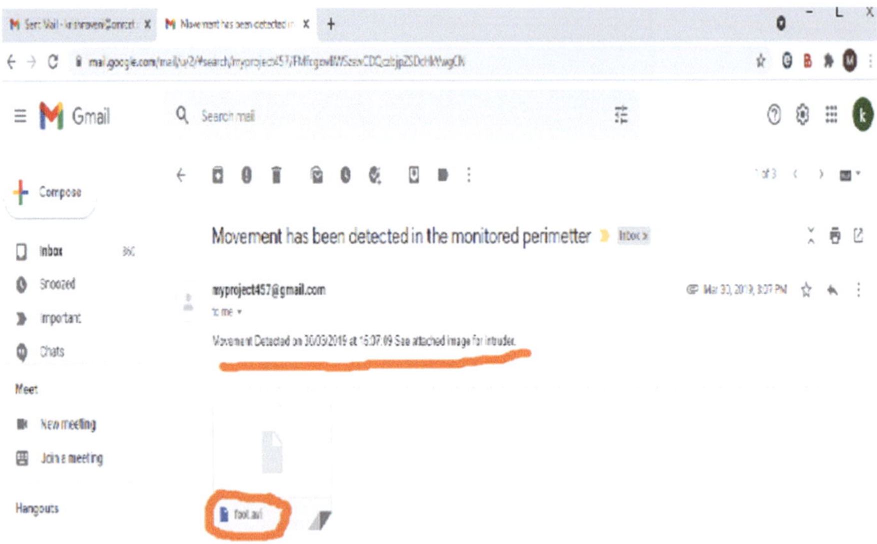

Fig. 19.8 Mail sent to the officials

If the movements are not detected, then the case structure will run the "default case" by indicating everything is normal in "string indicator", and the camera will monitor continuously (Fig. 19.8).

Conclusion

When infrared sensor detects any moving object in its premises, then it will alert the buzzer, the camera starts capturing a 30 s video, and it is stored in the given location with ".avi" extension, and then this ".avi" file has been sent to the official's registered mail account. Using myRIO portable device, it is easy to send the data in ".avi" form at a high accuracy.

References

1. Juhana, T., Angraini, V.G.: Design and implementation of smart home surveillance system. In: 10th International Conference on Telecommunication Systems Services and Applications (TSSA) (2016)
2. Harsha Vardhini, P.A., Harsha, M.S., Sai, P.N., Srikanth, P.: IoT based smart medicine assistive system for memory impairment patient. In: 2020 12th international conference on computational intelligence and communication networks (CICN), Bhimtal, India, pp. 182–186 (2020)

3. Babu, K.M.C., Harsha Vardhini, P.A.: Design and development of cost effective arduino based object sorting system. In: 2020 International Conference on Smart Electronics and Communication (ICOSEC), Trichy, India, pp. 913–918 (2020)
4. Harsha Vardhini, P.A., Ravinder, M., Srikanth, P., Supraja, M.: IoT based wireless data printing using raspberry Pi. J. Adv. Res. Dyn. Control Syst. **11**(4), 2141–2145 (2019)
5. Kumar, S., Swetha, S., Kiran, V.T., Johri, P.: IoT based smart home surveillance and automation. In: 2018 International Conference on Computing, Power and Communication Technologies (GUCON), Greater Noida, Uttar Pradesh, India (2018)
6. Saude, N., Vardhini, P.A.H.: IoT based smart baby cradle system using raspberry Pi B+. In: 2020 International Conference on Smart Innovations in Design, Environment, Management, Planning and Computing (ICSIDEMPC), pp. 273–278 (2020). https://doi.org/10.1109/ICSIDEMPC49020.2020.9299602
7. Kumar, P.: Design and implementation of smart home control using LabVIEW. In: Third International Conference on Advances in Electrical, Electronics, Information, Communication and Bio-Informatics (AEEICB), Chennai, pp. 10–12 (2017). https://doi.org/10.1109/AEEICB.2017.7972317
8. Vardhini, P.A.H., Koteswaramma, N., Babu, K.M.C.: IoT based raspberry pi crop vandalism prevention system. Int. J. Innov. Technol. Expl. Eng. **9**(1), 3188–3192 (2019)
9. Darrell, T., Demirdjian, D., Checka, N., Felzenszwalb, P.: Plan-view trajectory estimation with dense stereo background models. ICCV, pp. 628–635 (2001)
10. Harsha Vardhini, P.A., Ravinder, M., Srikanth Reddy, P., Supraja, M.: Power optimized arduino baggage tracking system with finger print authentication. J. Appl. Sci. Comput. J-ASC **6**(4), 3655–3660 (2019)
11. Kogut, G., Trivedi, M.: A wide area tracking system for vision sensor networks. In: 9th World Congress on Intelligent Transport Systems, Chicalgo, Illinois (2002)
12. Prakasam, V., Sandeep, P., Harsha Vardhini, P.A.: Snappy and video St ream edge detection using labview. Int. J. Adv. Sci. Technol. **28**(19), 197–203 (2019)
13. Shukla, A., Merugu, S., Jain, K.: A technical review on image super-resolution techniques. Lect. Notes Electr. Eng. **643**, 543–565 (2020)
14. Usha, S., Shaik, A.S.: Medical image analysis by pseudo color processing. Int. J. Res. Dev. Technol. **6**(4), 87–91, ISSN: 2349-3585 (2016)
15. Yick, J., Mukherjee, B., Ghosal, D.: Wireless sensor network survey. Comput. Netw. **52**, 2292–2330 (2008)
16. Yang, L., Yang, S.-H., Yao, F.: Safety and security of remote monitoring and control of intelligent home environments. In: Proceedings of IEEE International Conference on Systems Man and Cybernetics, pp. 1149–1153 (2007)

Chapter 20
Severity and Risk Predictions of Diabetes on COVID-19 Using Machine Learning Techniques

Vadthe Narasimha and M. Dhanalakshmi

Introduction

Chest X-ray images can be used to auto-detect, diagnose and classify diseases that affect the lungs and may cause cancer, TB, cardiovascular or COVID-19. This disease majorly affects the lungs, and it damages the organs with air pollution, tobacco, smoking, microtuxies' and air particles. As mentioned, N1H1, MARS-Cov, SARS-Cov-2 will affect the lungs and can be identified using chest X-ray along with that other diseases also. SARS-COVID-19 was identified by china from cats in 2003 and MARS-COVID-19 was identified by Saudi Arabia from camels in 2012. If COVID-19 is transferred from animal to human being, it is called SARS-Cov-2 that was identified by China in 2019 Wuhan [1–4]. This COVID-19 can damage the lungs very severely of those who are suffering from primary disease and old age people. Major diseases like TB, cancer, cardiovascular and COVID-19 are commonly diagnosed using chest X-ray (CXT) images. This paper focused on one of the most venerable diseases COVID-19 which causes lungs damage and can also be detected using chest X-ray images [5, 6]. Most of the researchers are focusing on early detection of COVID-19 with different types of approaches with best encourage. Researches proposed several domains with cumulative exiting results mentioned below [7, 8]. The convolution neural network (CNN) along with supportive activation function is used [11, 12].

Zhang et al. [3] received 18 layers of depth CNN that were pre-trained with the ImageNet dataset. The sigmoid was utilized as an enactment work, with the

V. Narasimha (✉)
JNTUH Research Scholar, Department of Computer Science and Engineering, CMR College of Engineering & Technology, Hyderabad, Telangana, India
e-mail: chinna.narasimha63@gmail.com

M. Dhanalakshmi
Department of Information Technology, JNTUH, Jagityla, Hyderabad, Telangana, India

V. Garcia Diaz and G. J. Rincón Aponte (eds.), *Confidential Computing*, Advanced Technologies and Societal Change,
https://doi.org/10.1007/978-981-19-3045-4_21

binary cross-entropy loss function. Wang et al. [4, 13] prepared and demonstrated the classification of chest X-ray images into three classes normal, COVID-19 and viral bacterial functions.

Severe acute respiratory syndrome corona SARS-Cov-2 is one of the vulnerable diseases in the world after WHO declared it as a pandemic in January 2020. According to statistics, the coronavirus has affected the world economically and physically. Eighty percentage of the COVID-19-affected people were cured without hospitalization and only 20% of the patients who were comorbidities and old age people were hospitalized [9, 10]. Comorbidity causes more severity when compared to normal patients. Comorbidity means patients suffering from primary or secondary disease and any other major disease. A comorbidity patient suffering from COVID-19 may be hospitalized or even face mortality. COVID-19 majorly is detected as mild, moderate and severe. A patient with severe disease from COVID-19 requires approximately 13 days of repository support, and the time duration may increase based on the individual's health. Patients with a growth rate of pf 25% to 35% of the affected region are considered to be with deadly cues and are admitted to intensive care units. COVID-19 clinical diagnosis depends on the various symptoms like fever (98% of cases), dry cough (75%), fatigue (45%), difficulty in breathing (55%), body pains (45%) and (ARDS) acute respiratory distress syndrome (20%). Severe cases may progress to multi-organ dysfunction along with the death of 2.5% if a patient suffers from diabetes. When we compare a normal COVID-19 patient to a diabetic COVID-19 patient, the severity of the COVID-19 disease is found more in a diabetic COVID-19 patient because of low immunity. The incubation period of the disease is for a longer duration in a diabetic when compared to a normal COVID-19 patient. This paper focuses on the diagnosis of the new disease COVID-19, the second section discusses methods, the third section deals with implementation, the fourth section follows results, and the last one is the conclusion. The dataset is collected from Kaggle.

Materials and Methods

COVID-19 chest X-ray image dataset and historical data of diabetes are collected from Kaggle internet. Using this dataset, the severity of COVID-19 and risk factors of diabetic patients affected with COVID-19 are implemented. Using COVID-19 chest X-ray images, we identified whether the patients are affected with COVID-19/normal pneumonia/bacterial pneumonia/viral pneumonia of COVID-19 using deep learning and machine learning techniques [14, 15].

Forward Propagation

Input layers are fed into an image in the form of numbers. This image pixels' value intensity may be denoted with numerical values. The hidden layer neurons apply few mathematical operations output values for generating the final prediction post the hidden layers input and output layers' values [16, 17].

Backward Propagation

After generating the output, we compare the output values with actual values. If the final output is close to the actual values, then the parameters are updated with the new output generated and are also applied for neural network algorithms [18].

Convolution Neural Network (CNN)

CNN or ConvNet is a deep learning algorithm, which can take an input image and classify it for recognition of objects from an image. CNN is used for a 16 layer deep neural network to recognize COVID-19 from chest X-ray images which can be followed by hidden layers, fully connected layers, active normalization functions and binary classification function sigmoid and dense layers for output. CNN is the first layer to recognize an image (Fig. 20.1).

ResNet

Residual neural network (RNN) is a specific type of neuron network that is used for deep residual learning for image recognition which replaces VGG16 layers in faster R-CNN with ResNet101. The ResNet may help in reducing the complexity of recognition that follows ReLU network (Fig. 20.2).

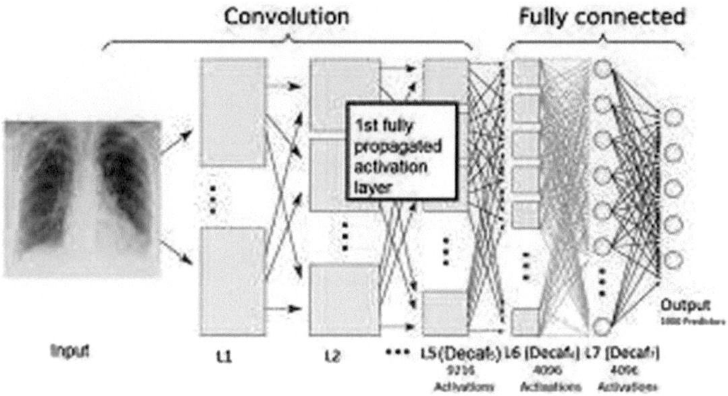

Fig. 20.1 CNN with medical training chest X-ray [2]

Fig. 20.2 ResNet CNN for image classification complex [5]

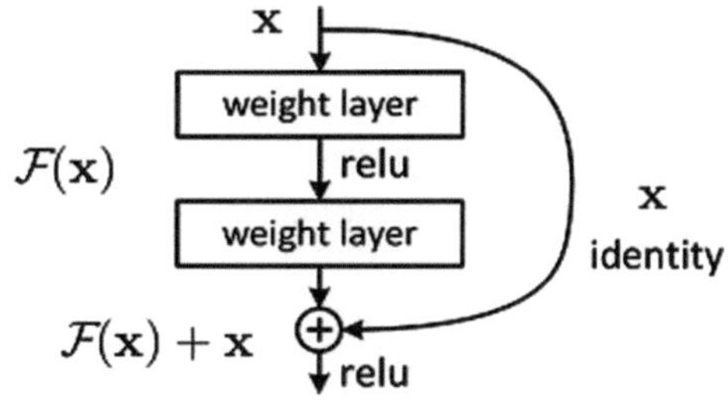

Pooling Layers

CNN pooling layer or dropdown layer is used to compress a number of network parameters. The pooling layer defines two types of max-pooling layers that can give the maximum parameter value from the matrix, and the average pooling layer can give the average network parameter values from the matrix in receptive fields of 2×2 matrix [19].

Activation Function

An activation function is a crucial part to design a neural network to get the input layer, hidden layer and output layer elements of the neural network. An activation function consists of sigmoid, ReLU, tanh function, softmax function, linear and nonlinear functions [20]. These functions wherever used for the output layer for binary classification either result in zero or 1. An activation function $f(x) = \max(0, x)$ is a rectified linear unit (ReLU) and generates a nonlinear activation map.

$$A = 1/\left(1 + e^{-x}\right) \tag{20.1}$$

$$Z = W^T \cdot X + b \tag{20.2}$$

where x is input, w is weight, b is constant that can be used for the matrix and define the weight and bias of the matrix with hidden layers.

$$Z = W^T \cdot X + b$$

$$Z = \begin{bmatrix} W_{11} & W_{21} & W_{31} & W_{41} \\ W_{12} & W_{22} & W_{32} & W42 \end{bmatrix} \begin{bmatrix} x_1 \\ x_2 \\ x_3 \\ x_4 \end{bmatrix} + \begin{bmatrix} b_1 \\ b_2 \end{bmatrix}$$

$$Z_{2\times 2} = \begin{bmatrix} W_{11}X_1 + W_{21}X_2 + W_{31}X_3 + W_{41}X_4 \\ W_{12}X_1 + W_{22}X_2 + W_{33}X_3 + W_{42}X_4 \end{bmatrix}$$

Now, there is one final step for the forward propagation to get linear and nonlinear functions output with best transactions.

Xception Network

CNN is a pre-trained model for image predictions on the ImageNet database with Python and Keras DL library especially in the Xception model [21]. ImageNet could be a project supposed to label and reason images manually. Within the field of DL and CNN, we will refer ImageNet as "ImageNet Large-scale Visual Recognition Challenge" briefly called ILSVRC. The main objective of this ImageNet project is to coach a model [22] which classifies an input image into 1000 separate classes. Xception network model support basically has 71 layers' deep convolution network [23]. It is divided into two parts: depth-wise Xception and point-wise Xception

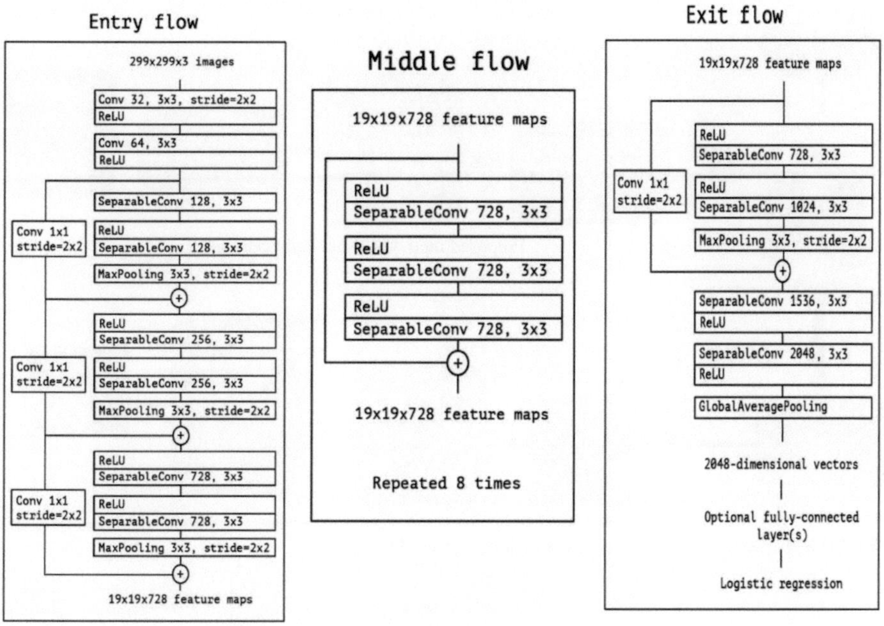

Fig. 20.3 Xception model implementation for COVID-19

network models. For the COVID-19 dataset, both implementations have been used [23] (Fig. 20.3).

Data Implementation

Convolution neural network (CNN) is used for the classification of the image along with the Xception model. Firstly, we can classify the image using CNN, ResNet51, InceptionV3 model to detect the SARS-Cov-2 and compare it with the newly implemented models. Here, three different dataset classes with a file size of dataset1 1648, dataset2 3564 and another dataset3 6566 are used. These datasets have been downloaded from Kaggle datasets1 and 2 and dataset3 from academia NIH [5]. To implement the data, first, we detected COVID-19 from CXT using DNN models; along with that, we have used the historical dataset of diabetic patients who are affected with the COVID-19 and long-standing diabetic patients. We observed that diabetic patients have very little immunity when compared to the other patients affected by COVID-19. So, COVID-19 severity is very high, and the risk factor of hospitalization causes mortality [9] (Figs. 20.4 and 20.5).

Then, at that point, the lungs opacities in the entirety of 30 lungs areas abstractly are assessed on chest X-beam. Every area has scored 0, 1 or 2 focuses relying upon the pulmonary parenchymal opacification included: 0%, 1-half, or 50–100%, separately

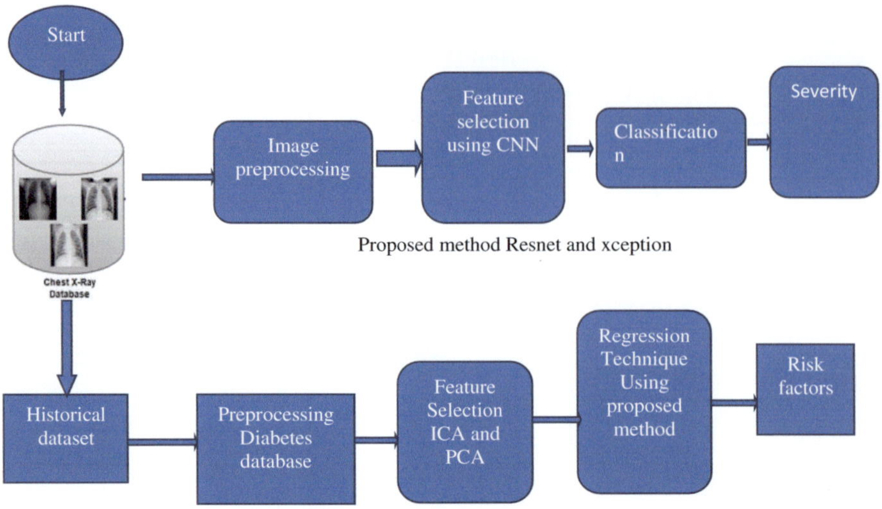

Fig. 20.4 Flow diagram of severity and risk factors detection

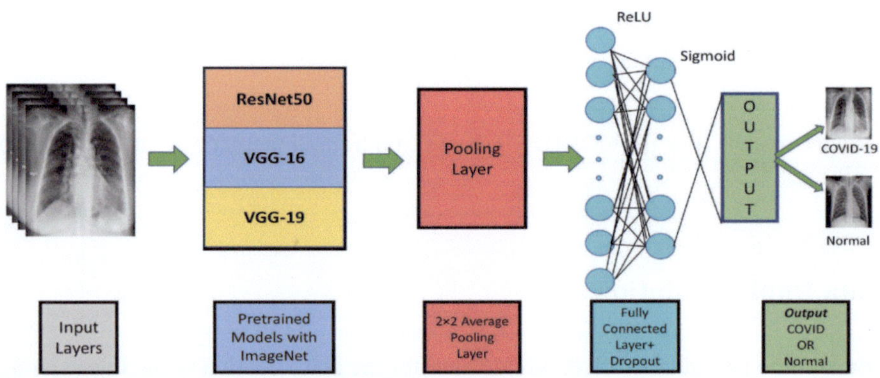

Fig. 20.5 COVID-19 detection from chest X-rays using DNN

[10]; the by and large CXT seriousness scores are characterized by the amount of the focuses scored for every one of the 30 lung fragment locals, which goes from 0 to 60 focuses. In the investigation, we have included 420 patients affirmed SARS-Co-2 contamination in RT-PCR (253 men and 147 women, aged 15 to 79 years, 84 cases with gentle and 18 cases with serious illness) transaction of DNA formation. The ideal CXT-SS edge for distinguishing extreme COVID-19 has 14.5 focuses with 84.3% affectability and 96% explicitness. The onlooker ICC for CXT-SS was observed to be phenomenal, with middle ICC 0.905 and mean ICC 0.966 for 420 patients [8].

Results and Discursion

This session discusses the resultant data for the trained 80% of the dataset with different cases of X-ray images, and the remaining 20% of the dataset were tested for the trained model to check whether they were working properly. The principle investigation was done to investigate the connection between the imaging appearances and clinical grouping of COVID-19. Every projection could be granted 0 to 4 focuses, contingent upon the level of the elaborate flap: 0 (0%), (1–25%), 2 (26-half), 3 (51–75%) or 4 (76–100%). The absolute seriousness score (TSS) was reached by adding focuses from every one of the five flaps [11]. The TSS cut-off for recognizing extreme basic sort was 7.5 with 82.6% affectability and 100% particularity [3]. The consistency of consequences of this strategy from two users shows a great loop with ICC—equivalent to 0.976 (95% certainty stretch 0.962–0.985).

Every projection has been defined by 5 levels of CXT image values contingent upon level of the elaborate flap: score 0—0% inclusion; score1—under 5% association; score2—5% to 25% contribution; score3—26% to 49% contribution; score4—half to 75% contribution; score5—more prominent than 75% contribution. Generally speaking, the CXT score amount of focus from every projection can reach from 0 to 25 points. After removing incentives for distinguishing extreme instances of COVID-19, the CXT score with affectability is 98.69%, 98.06%, 97.64% and 93.28%, separately (Fig. 20.6).

In a classification task, outcomes can be communicated in a particular framework called the "confusion matrix" (CM). In a binary classification task, CM consists of an accompanying data of occasions, and the quantity classifications, true positives (TP), are effectively perceived from the class of interest. True negatives (TN) that the number of cases accurately perceived that have a place with the false positives (FP) are occurrences that are allotted to the class revenue. However, we do not have to place it with false negatives (FN). [12]. The below images show all types of results.

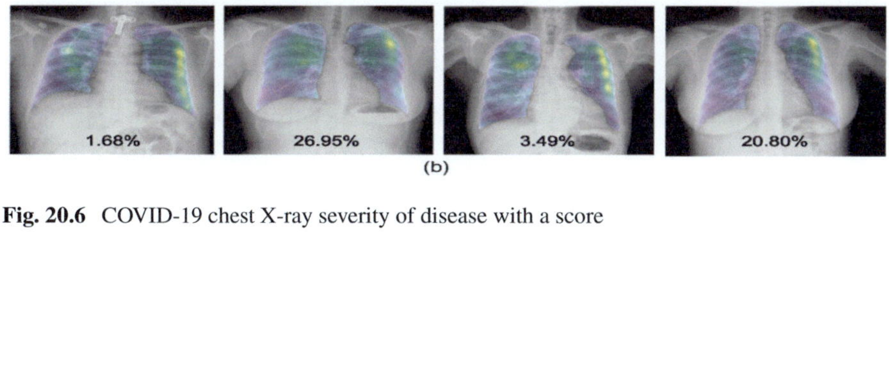

Fig. 20.6 COVID-19 chest X-ray severity of disease with a score

Table 20.1 Results with different models

Feature extraction	Classifier	accuracy
CNN	CNN	96.66
VGG16	SVM	90.53
VGG19	SVM	95.78
ResNet51	ImageNet	92.03
InceptionV3	SVM	97.75
Xception	**Depth-wise CNN**	**98.75**

Classification of common model performs a specific matrix measurement of values containing in CM; in binary classification, accuracy is most common for the overall effect of classifiers. Accuracy is defined as follows:

$$\text{Accuracy} = \frac{tp + tn}{tp + tn + fp + fn} \tag{20.3}$$

However, when working with common dataset imbalance, use different performance measures with accuracy, precision, recall and F1-score that are most common accuracy measurements used, and the curve of balanced and unbalanced values can be represented graphically. These measures are defined below:

$$\text{Precision} = \frac{tp}{tp + fp} \tag{20.4}$$

$$\text{Recall} = \frac{tp}{tp + fn} \tag{20.5}$$

$$F1 - \text{Score} = \frac{2tp}{2tp + tn + fp} = 2X\frac{\text{Pricision} \times \text{Recall}}{\text{Pricision} + \text{Recall}} \tag{20.6}$$

Here, the proportions of intermingling and training period for the proposed model and the pattern models are dense in Table 20.1. The best results highlighted are (Figs. 20.7, 20.8, 20.9, 20.10, 20.11, 20.12, 20.3, 20.14, 20.15 and Tables 20.2, 20.3):

Conclusion and Feature Scope

In this paper, machine learning and deep learning models are presented to perform prediction and classification of COVID-19 disease chest X-rays images, with a proposed method used for transferring learning techniques of Xception CNN pre-trained weight on ImageNet that has a new model for initialization. This procedure has 36 convolution layers, and for the yield, the stream contains a global average pooling layer, a dropout of 0.25 rates and a determined layer of 2 neurons along with

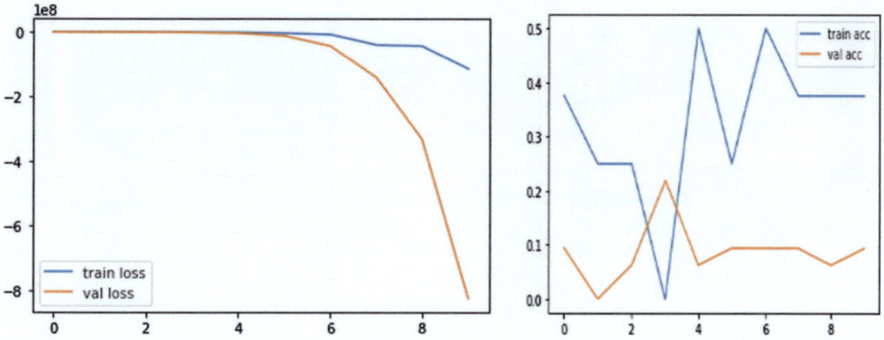

Fig. 20.7 Graphical representation of validation accuracy

Fig. 20.8 K-mean values

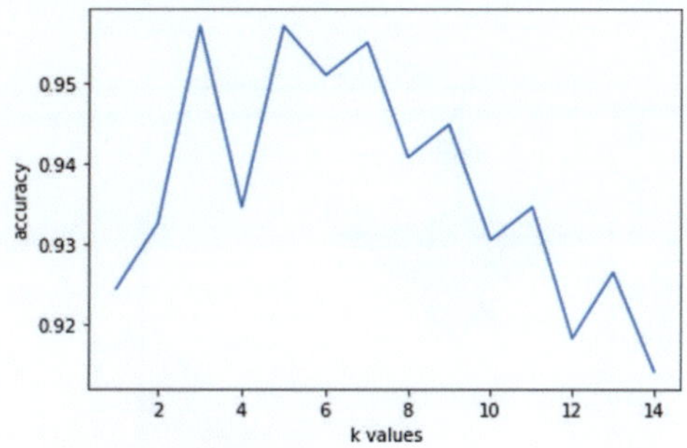

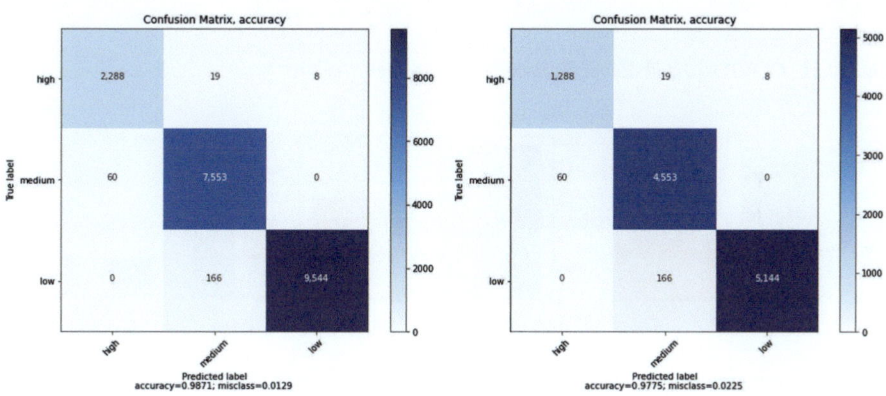

Fig. 20.9 Confusion matrix of COVID-19 with two different datasets

sigmoid limit initiation work. Xception NN provides the best results with **98.75% accuracy** and identifies the severity of COVID-19 patients using Co-RAD score, RT-PCR values of DNA and body temperature. These three diagnoses are used for identifying the severity of the patient.

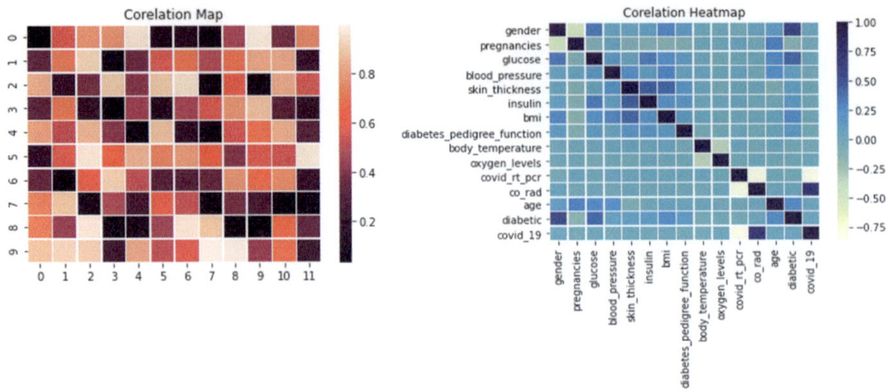

Fig. 20.10 Correlation map and heatmap of feature description

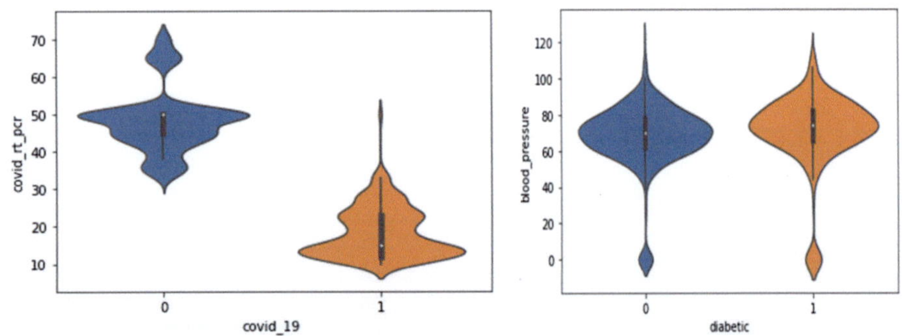

Fig. 20.11 COVID-19 and diabetic miscellaneous values

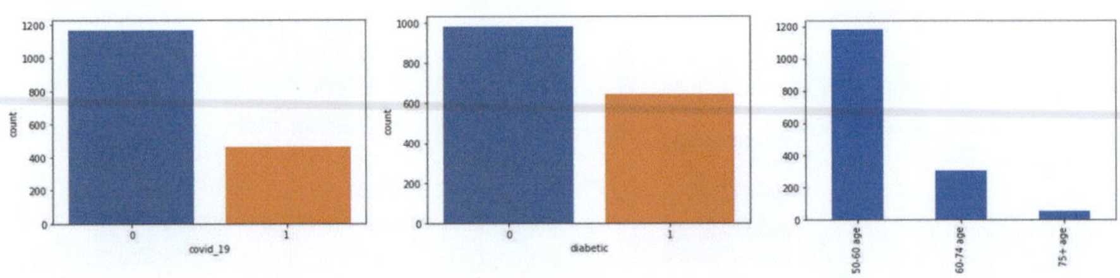

Fig. 20.12 Bar graph of patient age and individual defect of COVID-19 and diabetic miscellaneous values

In continuation of this work, we would like to consider the historical dataset results of COVID-19 disease and diabetic patients' contextual data along with image result set, and both pre-processing and data contextualization techniques involve with more effective and encouraging results for image wrapping. In addition, we would like to

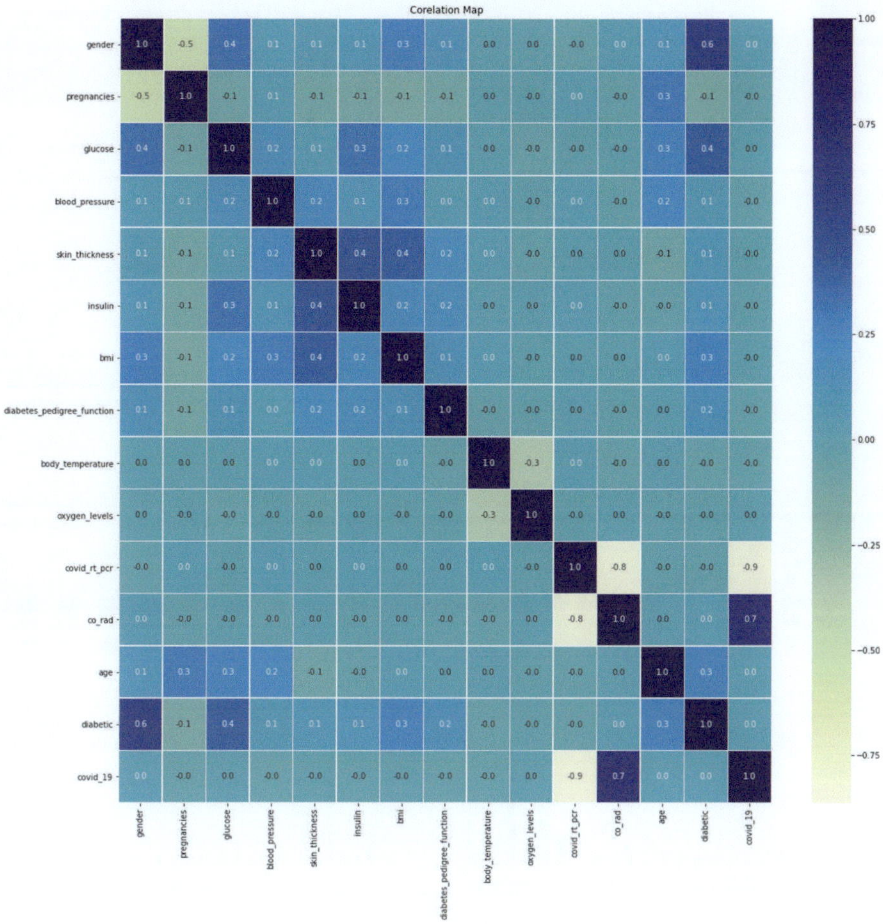

Fig. 20.13 Correlation map with accuracy

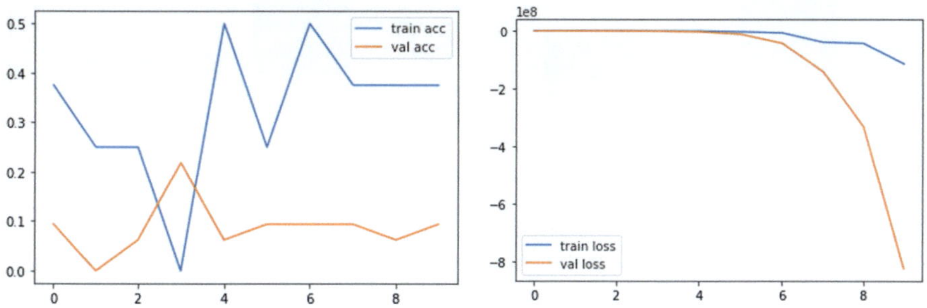

Fig. 20.14 Graphical representation of COVID-19 and diabetes severity with accuracy curves

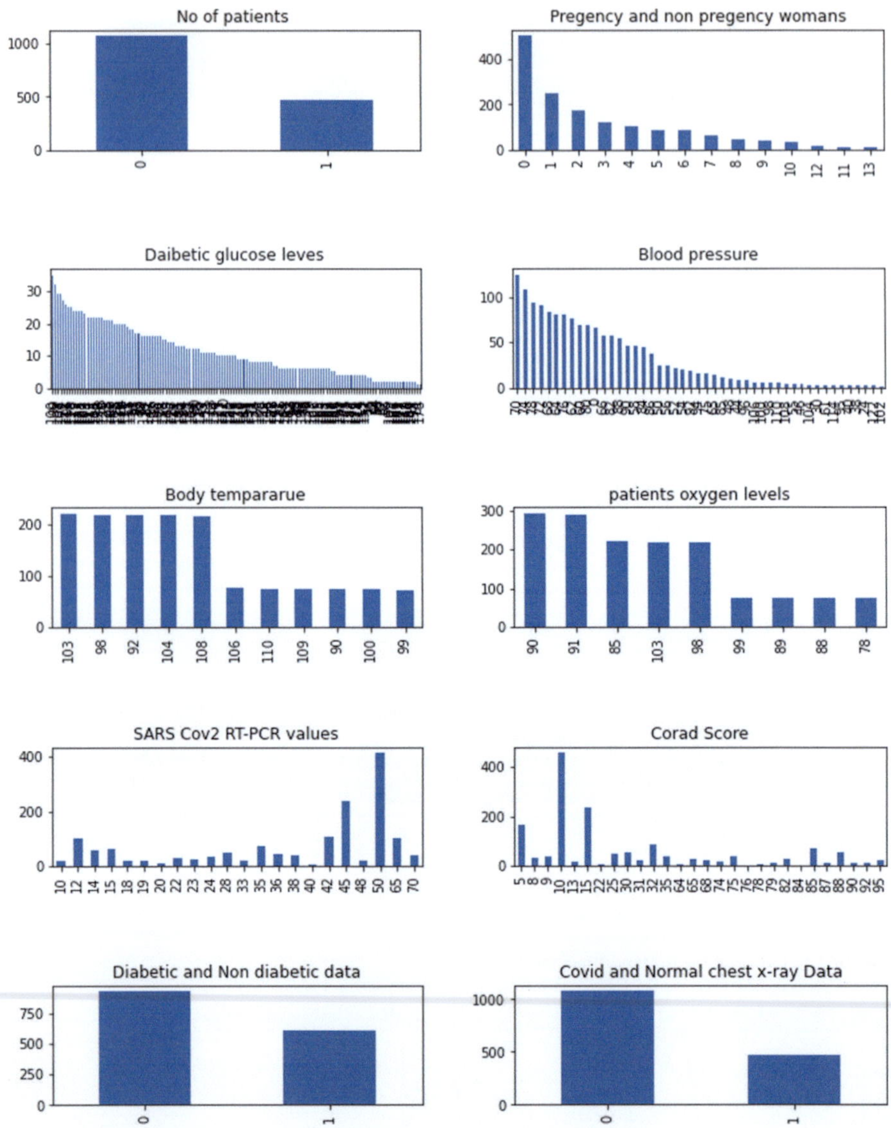

Fig. 20.15 Graphical representation of bar graph of dataset features

take several medical image datasets in order to create statistical analysis and risk factors.

Table 20.2 Confusion matrix measurement with different models

Model	Confusion matrix	Precision	Recall	F1-score	ROC curve AUC	Precision–recall AUC
CNN	174 45 45 398	**0.884**	**0.964**	**0.958**	**0.953**	**0.966**
VGG16	184 50 42 348	0.874	0.892	0.883	0.913	0.942
VGG19	156 78 4 386	0.832	0.940	0.904	0.946	0.955
ResNet51	188 46 5 365	0.891	0.927	0.927	0.913	0.935
DenseNet121	159 75 3 387	0.838	0.962	0.908	0.951	0.954
Xception	**149 65** **6 355**	**0.881**	**0.987**	**0.957**	**0.963**	**0.995**

Table 20.3 CNN model architecture

Model: "sequential"		
Layer (type)	Output shape	Param #
conv2d (Conv2D)	(None, 254, 254, 32)	896
conv2d_1 (Conv2D)	(None, 252, 252, 64)	18,496
max_pooling2d (MaxPooling2D)	(None, 126, 126, 64)	0
dropout (Dropout)	(None, 126, 126, 64)	0
conv2d_2 (Conv2D)	(None, 124, 124, 64)	36,928
max_pooling2d_1 (MaxPooling2	(None, 62, 62, 64)	0
dropout_1 (Dropout)	(None, 62, 62, 64)	0
conv2d_3 (Conv2D)	(None, 60, 60, 128)	73,856
max_pooling2d_2 (MaxPooling2	(None, 30, 30, 128)	0
dropout_2 (Dropout)	(None, 30, 30, 128)	0
flatten (Flatten)	(None, 115,200)	0
dense (Dense)	(None, 64)	7,372,864
dropout_3 (Dropout)	(None, 64)	0
dense_1 (Dense)	(None, 1)	65
Total params: 7,503,105		
Trainable params: 7,503,105		
Non-trainable params: 0		

References

1. Bae, S., et al.: Impact of cardiovascular disease and risk factors on fatal outcomes in patients with COVID-19 according to age: a systematic review and meta-analysis **107**(5), 373–380 (2021)
2. Boersma, H.E., et al.: Skin autofluorescence predicts new cardiovascular disease and mortality in people with type 2 diabetes **21**(1), 1–8 (2021)
3. Chiu, T.-T., et al.: The related risk factors of diabetic retinopathy in elderly patients with type 2 diabetes mellitus: a hospital-based cohort study in Taiwan **18**(1), 307 (2021)
4. Gregory, J.M., et al.: COVID-19 severity is tripled in the diabetes community: a prospective analysis of the pandemic's impact in Type 1 and Type 2 diabetes. **44**(2), 526–532 (2021)
5. Narasimha, V.: Education, diabetes with co-morbidities of Covid-19 using Covid-19 lungs images **12**(4), 1156–1164 (2021)
6. Gao, F., et al.: Obesity is a risk factor for greater COVID-19 severity **43**(7), e72–e74 (2020)
7. Narasimha, V., Dhanalakshmi, M.: A survey on context based medical image processing using machine learning 423–432 (2021)
8. de Sousa, P.M., et al.: COVID-19 classification in X-ray chest images using a new convolutional neural network: CNN-COVID 1–11 (2021)
9. Ahsan, M., et al.: COVID-19 detection from chest X-ray images using feature fusion and deep learning **21**(4), 1480 (2021)
10. Vignatelli, L., et al.: Risk of hospitalization and death for COVID-19 in people with Parkinson's disease or parkinsonism **36**(1), 1–10 (2021)
11. Narasimha, V., Satyanarayana, B., Krishnaiah, K.: Classification of knowledge based image using decision tree algorithm
12. Luján-García, J.E., et al.: A transfer learning method for pneumonia classification and visualization **10**(8), 2908 (2020)
13. Mangal, A. et al.: Covidaid: COVID-19 detection using chest x-ray. arXiv:2004.09803 (2020)
14. Wang, X., et al.: Chestx-ray8: hospital-scale chest x-ray database and benchmarks on weakly-supervised classification and localization of common thorax diseases. In: Proceedings IEEE Conference Computer Vision and Pattern Recognition, pp. 2097–2106 (2017)
15. Khan, A.I., Shah, J.L., Bhat, M.: Coronet: a deep neural network for detection and diagnosis of COVID-19 from chest x-ray images. Comput. Methods Programs Biomed. **196**, 105581 (2020)
16. Hall, L.O., et al.: Finding COVID-19 from chest x-rays using deep learning on a small dataset. arXiv:2004.02060 (2020)
17. Merugu, S., Reddy, M.C.S., Goyal, E., Piplani, L.: Text message classification using supervised machine learning algorithms. Lect. Notes Electr. Eng. **500**, 141–150 (2019)
18. Bukhari, S.U.K., et al.: The diagnostic evaluation of convolutional neural network (CNN) for the assessment of chest x-ray of patients infected with COVID-19. medRxiv (2020)
19. Devadasu, G., Sushama, M.: A novel multiple fault identification with fast fourier transform analysis. In: 1st International Conference on Emerging Trends in Engineering, Technology and Science, ICETETS (2016)
20. Sun, C., et al.: Revisiting unreasonable effectiveness of data in deep learning era. In: Proceeding IEEE International Conference Computer Vision, pp. 843–852 (2017)
21. Demirovic, D., Skejic, E., Šerifovic-Trbalic, A.: Performance of some image processing algorithms in tensorflow. In: 25th International Conference on System Signals and Image Processing IEEE, pp. 1–4 (2018)
22. He, K., et al.: Deep residual learning for image recognition. In: Proceedings of IEEE Conference Computer Vision and Pattern Recognition, pp. 770–778 (2016)
23. Deng, J., et al.: Imagenet: a large-scale hierarchical image database, pp. 248–255. IEEE Comput. Vision Pattern Recognit., IEEE (2009)

Chapter 21
Detection of Cyber Threats in Application Platforms

Krishnaveni Kommuri, Vamsee Krishna Allam, Ritika Allam, Vanapalli Geethika, and Boyapati Vyshnavi

Introduction

Cybersecurity governs the data up-strings. Commercial rate of social websites includes the motive that they need an approach to make exact assets for dereliction. Deceitful designers use portable stockpile purpose of ransom ware. Upon awareness produced by their clients also effects online services and gives better field to impact the citizens thinking. Performance of electronic devices appeal in trade depends upon a search rank such as Google Play. Mostly, low-ranking apps are fixed and also produce more remuneration from promotions.

A Pseudo Clique Finder program is proposed which tracks the behavior of deceiver recruit to analysis/download/rate an application that is such as to advertise those evaluation/rating in less interval of duration or download the app using symmetric key. The server needs to permit the key to access the secret key as referred to Table. 21.1. PCF takes the input as set of downloads of an app and limit value ($u = 3$).Whenever the threshold limit exceeds by downloading the application more than the u times, the user is said to be fraudster as showed in Fig. 21.5.

There is a most usual view that high-ranking users support the latest applications to bring out high search grades. The best aim for users is to advertise their own implementations to create an App Search Optimization submarket as shown in Fig. 21.1. A process advancement approach will combine implementation, examine cases and enlargement plans to an efficiently defined number of steps that are key to program development establishment. The analyst works on this issue of finding mislead violation to classify the data. This analyst gives detailed report about the outcome and analysis of it using artificial data as shown in Fig. 21.2.

K. Kommuri (✉) · V. K. Allam · R. Allam · V. Geethika · B. Vyshnavi
Department of ECM, Koneru Lakshmaiah Education Foundation, Guntur, AP, India
e-mail: krishnavenikommuri@kluniversity.in

V. Garcia Diaz and G. J. Rincón Aponte (eds.), *Confidential Computing*, Advanced Technologies and Societal Change,
https://doi.org/10.1007/978-981-19-3045-4_22

Table 21.1 Symmetric key permission

ID	User name	File name	Permission
1	Ritika	WhatsApp	Permitted
2	Geethika	Facebook	Permit
3	Vyshnavi	Easy Task Lazy Swipe	Permitted

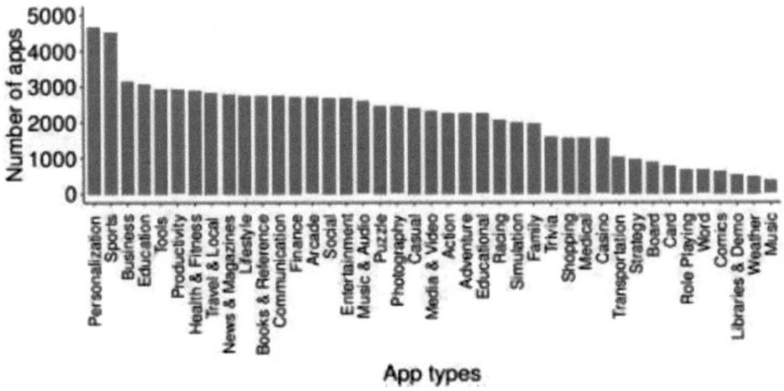

Fig. 21.1 App types (to create an App Search Optimization (ASO) submarket)

Fig. 21.2 Average rating of app SVS number of apps

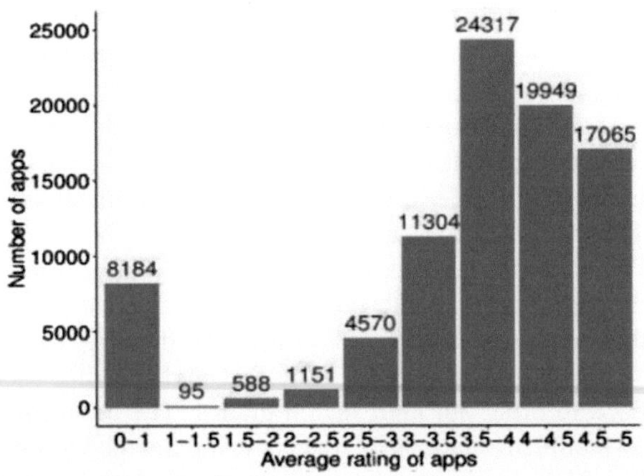

Literature Survey

Kotkar and Sucharita [1] ML' programs can be effectively employed for modeling and developing autonomous vehicle systems. SVC and Naïve Bayes algorithms have been found to be quite suitable for modeling autonomous vehicle systems. Naïve Bayes algorithm has been proved to be the best choice due to increased accuracy, less prediction time and training time [2]. DDOS attack detection models generate a large set of patterns (or signatures) in which most of them are in accurate due to high false alarm rate. Traditional packet correlation approaches need several network packets beside knowledgeable information to forestall complicated DDOS attacks. Also, detection and anticipate on of dynamic DDOS attacks are difficult in the real-time distributed LAN/WLAN networks [3]. An examiner present in IDS is used to carry out deep packet' examination. Somu et al. [4] cyber threats are increasing, and one of the common aspects of all attacks has a commonality, which is a malware [5]. In learning dynamic malware analysis, its future scope can be further extended to static malware analysis where we completely try to unpack and learn the working principle of a malware, further approach can be to reverse engineer it, and from this paper, user gets a proper understanding about dynamic analysis approach [6]. Dynamic analysis of a malware can predict the behavior of malware, and we can plan the strategy to take down the malware [7]. In dynamic analysis as soon as various organization can benefit from dynamic analysis of malware forensics, because at certain point in life of a network administrator, he/she has to encounter the type of malware affected and should take necessary steps to counter it [8]. This paper explains various spam located in various actors in everyday life.

Search Security Shield Tracking Technology

Basically, a shield jaunt software would permit workplace employees to put digital inspect by the side of a defend scouting [9–12]. Inspect points are utterly organized through the workplace people with agile monitoring gadget, and it prompts robotically when a protector comes to a definite place throughout their scouting. Like the forte is examined, for that reason, permit the gathering of unique facts. Later getting into inspection, protector can scan barcodes and QR codes with their app. The laptop routinely traces when a soldier testinata precise spot, defines the information is efficiently conveyed and even though the protector's laptop temporarily based on moment's will lose service [13–15].

Proposed Methodology

The suggested task develops [16–18] proving the center view outlines of client survey associations. The manufacture PCF, which is a compelling heuristic for perceive in no time restricted, center vision foremost pops set up by the commentators when the with impressively crossing center view occasions across brief time frame entry ways. The utilizing momentary components of remark time survey to characterize the suspect investigation floods bought by applications. As shown that a common need to post at any rate certain criticism to counterbalance for an antagonistic audit, for an appeal which has a result [19]. The regularly order appeals with "shaky" means examination, positioning and establishment, [20, 21] just as applications with the slopes for asking acknowledgment. This work depends on then the thought that they

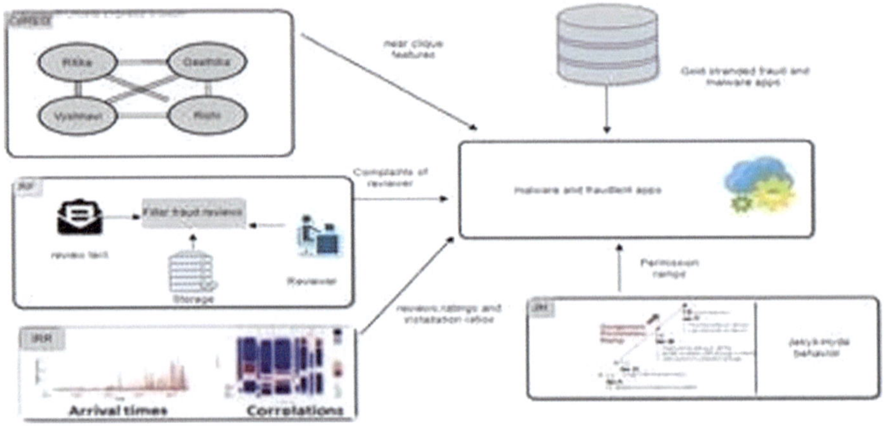

Fig. 21.3 Behavior of apps (different frames which are used to detect thousands of malicious reviews and ratings)

ill-conceived and deceitful practices on application stages leaving the conspicuous signs aside.

Misrepresentation rank finder Fig. 21.3 carries out over 97% exactness in the discovery of malignant and the asking programs, and the 95% over the rightness the distinguishing proof of malware and generous usage.

Authentic exploit-influenced clients will record unfavorable the remarks that have information. Fraud rank detector espies thousands of spiteful applications as showed in Fig. 21.3. In contrast to existing arrangements, the proposed frame work builds up its exploration onto the discovering which illicit and destructive exercises on combine the versatile areas left away admonition actions which is a suggested framework that reveals these terrible works by choosing these tracks.

Interface Architecture

1. Login or register
2. Outline your account.
3. Explore applications and recommendations, by likes, given ranking 5, 4, 3 or 2
4. Appeal for symmetric key for application to download.
5. To download application, symmetric key status should be permitted as shown in Table. 21.1.
6. Steps executed by users, developers on apps as referred in Fig. 21.4.
7. PCF algorithm pseudocode

    i. **Start:** for d:=0’ ‘d<days.Size();d++
    ii. Graph’‘PC:=newGraph();
    iii. ‘bestNearClique(PC, days[d])’

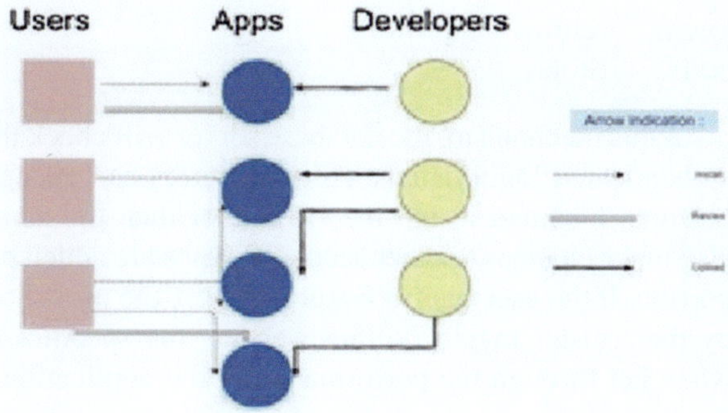

Fig. 21.4 System architecture

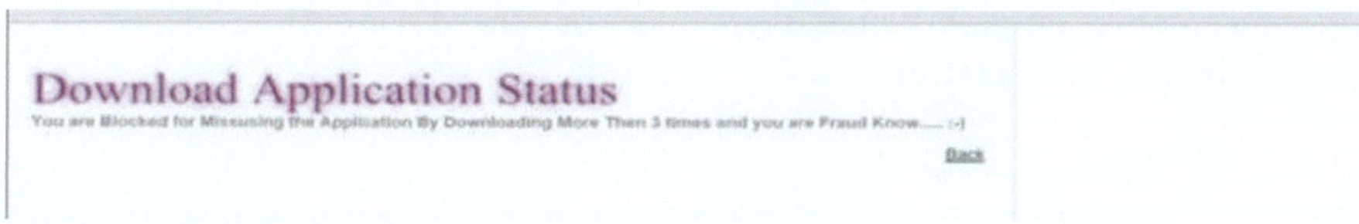

Fig. 21.5 Fraud detection status (status after blocking the user when he/she exceeds threshold limit)

iv. ‘c:=1;n:=PC.size()’;
v. fornd:=d+1;d<days.size()&c=1;d++’
vi. bestNearClique(PC,days[nd])’
vii. c:=(PC.size()>n);endfor’
viii. if(PC.size()>2)’
ix. allCliques:=allCliques.add(PC);endfor’
x. return.
xi. functionbestNearClique(GraphPC,Setrevs)’
xii. if(PC.size()=0)’
xiii. forroot:=0;root<revs.size();root++’
xiv. GraphcandClique:=newGraph();’
xv. candClique.addNode(revs[root].getUser());’
xvi. docandNode:=getMaxDensityGain(revs);’
xvii. if(density(candClique{candNode}))’
xviii. candClique.addNode(candNode);’
xix. while(candNode!=null);’
xx. if(candClique.density()>maxRho)’
xxi. maxRho:=candClique.density();’
xxii. PC:=candClique;endfor’
xxiii. elseif(PC.size()>0)’
xxiv. docandNode:=getMaxDensityGain(revs);’
xxv. if(density(candClique[candNode)))’
xxvi. PC.addNode(candNode);’
xxvii. while(candNode!=null);’

xxviii. return.
xxix. Stop

In this mechanism, the database server will check the current devices which have authenticated login details. Administrators can change privacy settings, access all personal files and install the OS. Other than the administrators, management can also make changes to other accounts. Apps are added by the administrator during the portion. If the user tries to feature the app, the user should type application followed by the register toggle. Within website, the specifics are going to be found. Once when get through the performance of the application, then view the details of the implementation device, consumers, the portal category, depiction. When admin tries for defraud, those details will also effect on performance ID, Ethernet address (fraud), app name, mobile, program name. It repudiates or concedes user from connecting to another computer and networks.

Results and Discussion

To detect spam feedback, the paper introduces the classification criteria of a series of opinion spam detection based on the spammer's behavioral characteristics. Related reviews and related feedback from all reviews are then suggested to be remembered by two algorithms. This is highly perfect and had features of an effect: Increased accuracy in classifying fraud applications, the fraud classification detector achieves an accuracy of over 97%, thus more than 95 million validity in identifying fraud and harmless applications. Actual impact: prove malware and provocations. Play Store finds many fake applications. Consequently, to spot and recognize more kinds of malware, it is imperative to assemble a lot of performance that add other wicked parts. Even though electronic device malware recognition stays tremendous challenge (Fig. 21.5).

Conclusion

This paper is proposed to acknowledge the malicious behavior such as misrepresentation in the application and make their rating/search rank zero. It also improves the performance by detecting the malware application and placing the apps with authorization. Using the list makes an interesting improvement in insurance if the malicious applications are hindered in the underlying download and evaluation phase.

References

1. Kotkar, V.A., Sucharita, V.: A comparative analysis of machine learning based anomaly detection techniques in video surveillance. J. Eng. Appl. Sci. **12**(Specialissue12), 9376–9381 (2017)
2. Arshad, M., Hussain, M.A.: A novel probabilistic based DDOS attack detection and prevention framework for dynamic LAN/WLAN networks. J. Adv. Res. Dyn. Cont. Syst. **9**(2), 272–286 (2017)
3. Rao, K.V.S.N.R., Battula, S.K., Krishna, T.L.S.R.: A smart heuristic scanner for an intrusion, detection system using two-stage machine learning techniques. Int. J. Adv. Intell. Parad. **9**(43987), 519–529 (2017)
4. Somu, V., Kamesh, D.B.K., Sastry, J.K.R., Sitara, S.N.M.: Snort rule detection for countering in network attacks. Adv. Intell. Syst. Comput. **515**, 573–583 (2017)
5. Krishna Anne, V.P., Rajasekhara Rao, K.: Standards and analysis of intrusion detection-based system: a comparative study. Ponte **73**(2), 87–97 (2017)
6. Vara Prasad, P.V., Sowmya, N., Rajasekhar, R.K., Jayant Bala, P.: Introduction to dynamic malware analysis for cyber intelligence and forensics. Int. J. Mech. Eng. Technol. **9**(1), 10–21 (2018)
7. Divya, P., Saikiran, M., Nagarjuna, K., Kiran, K.V.D., Phani Krishna, C.V.: Correlation analysis. Int. J. Eng. Technol. (UAE) **7**, 230–236 (2018)
8. Bhavika, S., Prema Sindhuri, B., Bhavana, G.: Spam detection using semantic web in mail services. Int. J. Eng. Technol. (UAE) **7**(2), 44–47 (2018)
9. Swapna Goud, N., Mathur, A.: A certain investigations on web security threats and phishing website detection techniques. Int. J. Adv. Sci. Technol. **28**(16), 871–879 (2019)
10. Arshiya, M., Srikanth, V.: Advanced snort driven collaborative framework for DDOS attack detection in network classification. Int. J. Innov. Technol. Expl. Eng. **8**(7), 1300–1304 (2019)
11. Harikanth, M., Rajarajeswari, P.: Malicious event detection using ELK stack through cyber threat intelligence. Int. J. Innov. Technol. Expl. Eng. **8**(7), 882–886 (2019)
12. Ravinder Rao, P., Sucharita, V.: A framework to automate cloud based service attacks detection and prevention. Int. J. Adv. Comput. Sci. Appl. **10**(2), 241–250 (2019)
13. Manjula Josephine, B., Raja Shekar, K.V., Meghana, G., Rama Rao K.V.S.N., Rajesh, C.: Forensic analysis of social media apps. Int. J. Innov. Technol. Expl. Eng. **9**(1), 2302–2305 (2019)
14. Kolla, N.P., Sastry, J.K.R., Chandra Prakash, V., Onteru, S.K., Pinninti, Y.S.: Assessing quality of websites based on multimedia content. Int. J. Eng. Technol. (UAE) **7**(0), 1040–1044 (2018)
15. 'SaiHarika, T., Madhuri, N., Paraphrased, P.V.V.: Detection, prevention and mitigation of blackhole attack for MANET. Int. J. Recent Technol. Eng. **8**(1), 381–386 (2019)
16. Quinonez, R., Giraldo, J., Salazar, L., Bauman, E., Cardenas, A., Lin, Z.: Securing autonomous vehicles with a robust physics-based anomaly detector. In: 29th USENIX Security Symposium (USENIX Security 20) (2020)
17. Vedavalli, P., Krishnaveni, K., Sastry, J.K.R.: Securing data transmission using DES for smart home monitoring system. Int. J. Innov. Technol. Expl. Eng. (IJITEE) **8**(7) (2019), ISSN: 2278–3075
18. Krishnaveni, K., Kolluru, V.R.: Prototype development of CAQSS health care system with MQTT protocol by using Atmega328. In: 2020 International Conference on Artificial Intelligence and Signal Processing (AISP). IEEE (2020)
19. Sridevi. S., Kompalli, V.S.: An overview on prediction of plant leaves disease using image processing techniques. In: IOP Conference Series: Materials Science and Engineering, vol. 981, no. 2, p. 022024. IOP Publishing (2020)
20. Nayak, S.C., Misra, B.B.: Estimating stock closing indices using a GA-weighted condensed polynomial neural network. Financ. Innov. **4**(1) 2018
21. Krishnaveni, K., Kolluru, V.R.: Real time implementation and comparison of ESP8266 vs. MSP430F2618 QoS characteristics for embedded and IoT applications. Int. J. Adv. Comput. Sci. Appl. (IJACSA) **11**(9) (2020), ISSN:2156–5570

MIX
Papier aus verantwortungsvollen Quellen
Paper from responsible sources
FSC® C105338

If you have any concerns about our products,
you can contact us on
ProductSafety@springernature.com

In case Publisher is established outside the EU,
the EU authorized representative is:
Springer Nature Customer Service Center GmbH
Europaplatz 3, 69115 Heidelberg, Germany

Printed by Libri Plureos GmbH
in Hamburg, Germany